This book is to be returned on or before the last date stamped below

Longman Handbooks in Agriculture

Animal production and health
Series editors

C. T. Whittemore
R. J. Thomas
J. H. D. Prescott

Whittemore: *Lactation of the dairy cow*
Whittemore: *Pig production: the scientific and practical principles*
Speedy: *Sheep production: science into practice*
Hunter: *Reproduction of farm animals*

Reproduction of farm animals

R. H. F. Hunter

School of Agriculture,
University of Edinburgh

 Longman London and New York

Longman Group Limited
Longman House
Burnt Mill, Harlow, Essex, UK

Published in the United States of America
by Longman Inc., New York

© Longman Group Limited 1982

First Published 1982

Printed in Hong Kong
by Astros Printing Ltd.

British Library Cataloguing in Publication Data

Hunter, R. H. F.
    Reproduction of farm animals.
    (Longman handbooks in agriculture)
    1. Reproduction 2. Stock and stock-breeding
    I. Title
    636.089'26   SF105
    ISBN 0-582-45085-3

Library of Congress Cataloging in Publication Data

Hunter, R. H. F.
    Reproduction of farm animals.
    (Longman handbooks in agriculture)
    Bibliography: p.
    Includes index.
    1. Livestock – Reproduction. 2. Veterinary physiology.
    I. Title
    II. Series
    SF 768.H83   636.089'26   81-19335
                                    AACR2

Contents

Preface

This book has been written principally for first and second year undergraduates in Agriculture or the Biological Sciences, for students in Colleges of Agriculture and Farm Institutes, and for farmers, advisers and managers with some biological training.

As to the contents of the six chapters, a qualifying statement should be offered at the beginning. Reproductive physiology is concerned with the science of reproduction – that is, seeking to understand the manner in which animals are able to breed successfully, thereby trying to explain situations in which breeding is inefficient or has failed completely. The manipulation of repdoduction – a field that is frequently termed *reproductive technology* – is an important and rapidly growing branch of applied science, and has much to contribute to the increased productivity of farm animals. Even so, reproductive technology is only readily understood and satisfactorily applied when the underlying science has been mastered. The format of the book therefore follows the suggested sequence with a description of reproductive processes in the first four chapters, a brief consideration of reproductive inefficiency in Chapter 5, and a detailed account of reproductive technology in Chapter 6. Although the chapters can be perused as separate entities, it is the author's hope that readers will persevere with the relatively detailed Chapters 1 and 2 before moving on to more familiar territory in later sections of the book.

Acknowledgements for assistance in the preparation of the

text could be lengthy as numerous individuals have helped in a variety of ways, which include the associated teaching and research activities. Limitations of space prevent me from thanking all these colleagues in Edinburgh, but a special debt of gratitude is due to Mrs Pat Gallie for the time and trouble she took in typing the manuscript. The series editor, Dr Colin Whittemore, offered useful comments on an early draft of the manuscript, as did Miss Suzanne Crabtree of the Royal (Dick) School of Veterinary Studies. My technical assistants, Robert Nichol and Robert King, helped in diverse ways, and Mr Gordon Finnie prepared many of the photographic plates. Other illustrative material was kindly donated by Dr C. Polge, Director of the Animal Research Station in Cambridge, and by Dr G Seidel of the Embryo Transfer Laboratory, Colorado State University. Gratitude is also expressed to the staff of the Longman Group Limited for care and assistance in seeing the work through the press.

Finally, I must record my best thanks to my wife and family for endless help in preparation and checking of the manuscript, and for preventing me from reading other people's books in order that I should persevere with writing my own!

R.H.F.H.
Edinburgh
August, 1981

The female, puberty and oestrous cycles

1

In order to understand control of the reproductive functions, it is necessary to appreciate the relationship between the pituitary gland, hypothalamus and higher centres of the brain. Hormones of the pituitary gland regulate reproductive functions, and the overall integration of pituitary activity with the animal's behaviour and performance is achieved through the hypothalamus. In brief, monitoring of the animal's external and internal environment occurs in the brain due to a complete spectrum of nervous inputs, from where appropriate information is transmitted to the hypothalamus, and the endocrine sequence of responses proceeds from this level downwards. This may sound a far cry from understanding or indeed modifying reproductive processes in farm animals, but in fact preparations of hypothalamic and pituitary hormones have been used for precisely the latter purpose – especially to stimulate activity of the ovaries.

Hypothalamus and pituitary gland

As can be seen from Fig. 1.1, the hypothalamus is situated in the ventral region of the fore-brain and the pituitary gland projects beneath it on a stalk into a bony cavity. The pituitary is composed of two distinct portions, the anterior and posterior, and although these may appear closely related, they have been derived embryologically from completely different tissues; this derivation is reflected in their function. The

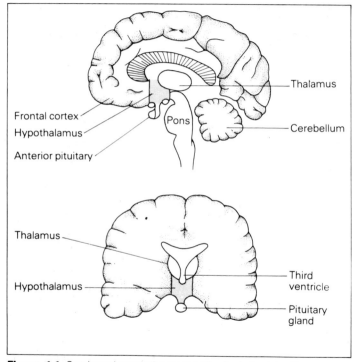

Frontal cortex

Hypothalamus

Anterior pituitary

Pons

Thalamus

Cerebellum

Thalamus

Hypothalamus

Third ventricle

Pituitary gland

Figure 1.1 Sections through the mammalian brain to show the anatomical relationship between the hypothalamus and the pituitary gland

anterior pituitary comprises specialised ectodermal tissue from an upgrowth of the roof of the embryological mouth, whereas the posterior pituitary is of nervous tissue origin formed by a downgrowth from the floor of the fore-brain. Because the posterior pituitary retains its nervous connections with the brain, there is no problem in controlling its function: nerve fibres pass directly from the hypothalamus into this portion of the gland. By contrast, the anterior pituitary is not innervated, and the question therefore arises as to the manner in which its secretion of hormones is controlled by the brain.

The link is to be found in the basal region of the hypothalamus. This contains a complex series of capillary loops which pass down the surface of the pituitary stalk and vascularise the anterior portion of the gland (Fig. 1.2). Nerve endings in the hypothalamus terminate in contact with the capillary loops and secrete specific products (*releasing hormones*) into the blood. These hormones are then transported through the capillary portal system to the anterior portion of the gland where they regulate synthesis and secretion of trophic hormones. Control of the anterior pituitary is therefore indirect by neurosecretion of releasing hormones, whereas control of the posterior pituitary is by direct innervation. In fact, the small peptide molecules that act on the anterior pituitary correspond closely with the hormones of the posterior pituitary which themselves are synthesised in the hypothalamus and only stored in the posterior gland.

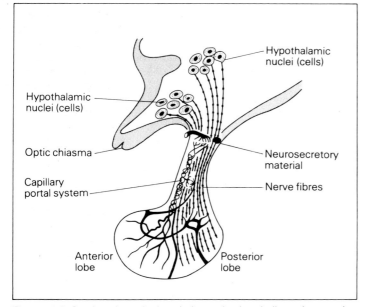

Figure 1.2 Section through the pituitary gland to indicate its anterior and posterior lobes and their functional relationships with the hypothalamus. Neurosecretory material from cells in the hypothalamus is transmitted *directly* down the nerve fibres to the posterior lobe whereas similar material reaches the anterior lobe *indirectly* via a capillary portal system. The hypothalamic factors regulating anterior pituitary function are termed 'releasing hormones'

Table 1.1 The six classical hormones of the anterior pituitary gland and the corresponding hypothalamic releasing hormones

Pituitary trophic hormone	Abbreviation	Releasing hormone
Growth hormone (somatotrophin)	GH (or STH)	GH-RH
Thyrotrophic hormone	TSH	TSH-RH
Adrenocorticotrophic hormone	ACTH	ACTH-RH
Follicle stimulating hormone	FSH ⎫	
Luteinising hormone	LH ⎬	Gn-RH*
Prolactin†	P	⎰ P-RH ⎱ P-IH

*Gn-RH is a hypothalamic hormone that releases FSH and LH.
†Prolactin is under the hypothalamic control of a releasing hormone and an inhibiting hormone.

The anterior pituitary gland secretes six classical hormones (Table 1.1), of which three are concerned with general body functions (growth hormone, thyrotrophic hormone and adrenocorticotrophic hormone) and three are directly concerned with reproductive events (follicle stimulating hormone, luteinising hormone and prolactin). Specific hypothalamic releasing hormones regulate secretion of these trophic hormones (Table 1.1). Although FSH and LH are known as gonadotrophic hormones and can be regarded as acting as a *gonadotrophic complex*, there is increasing evidence that prolactin should be regarded as part of this complex in many reproductive situations. In the male, FSH is primarily involved in stimulating spermatogenesis while LH is

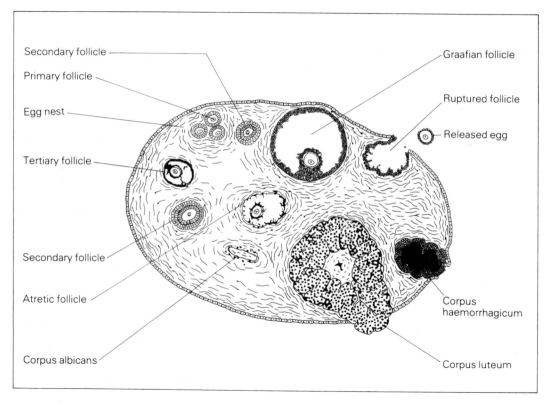

Secondary follicle

Primary follicle

Egg nest

Tertiary follicle

Secondary follicle

Atretic follicle

Corpus albicans

Graafian follicle

Ruptured follicle

Released egg

Corpus haemorrhagicum

Corpus luteum

Figure 1.3 Simplified section through a mammalian ovary after puberty to illustrate growth of Graafian follicles, release of the egg at ovulation, and subsequent formation of the corpus luteum. Follicular growth and ovulation are under the influence of FSH, LH and prolactin whilst formation of the corpus luteum is regulated by LH and prolactin

more concerned with regulation of male sex hormone (androgen) production by the interstitial tissue of the testes (see Ch. 2, p. 26). Prolactin also appears to contribute to the support of spermatogenesis. In the female, these gonadotrophic hormones are again involved in cellular and endocrine regulation, but here are secreted in cyclic fashion. FSH, in conjunction with prolactin, influences growth and maturation of ovarian follicles and stimulates their production of oestrogens. Ovulation of follicles with release of the egg requires a peak of LH secretion, after which the gonadotrophins regulate progesterone secretion by the corpus luteum (Fig. 1.3). By definition, therefore, the gonadotrophic hormones stimulate function of the testes and ovaries, whereas the gonadal hormones (i.e. the androgens, oestrogens and progesterone) act on the tissues of the reproductive tract. These gonadal steroids also act in a feedback loop to regulate hypothalamic and pituitary activity (Fig. 1.4), and at the level of the brain they influence sexual, aggressive and other forms of behaviour.

One of the objectives of reproductive technology in farm animals is to simulate or modify the secretion of gonadotrophic hormones so that ovarian activity in particular may be controlled. This would have considerable attraction if we could regulate the number of eggs shed at mating in cattle, or overcome the influence of seasonal breeding in sheep, by maintaining the ovaries in an active state. As discussed in Chapter 6, these objectives have not been attained with any precision, and there are several explanations. Possibly the most important is that the hypothalamus stimulates the anterior pituitary by releasing hormones in pulses or surges every one or two minutes, and this pulsatile form of hormone release or its effects are clearly very difficult to mimic in treatment on the farm. Second, the response to injection of gonadotrophic hormones involves a feedback action of the gonadal steroids as noted above, and this interferes with secretion of endogenous hormones and the precision of response. A third reason is that preparations of hormones used for injection are invariably gonadotrophins from a non-specific source – in fact, foreign proteins – and there is no reason to suppose that accurate dose-response stimulation will follow.

The posterior pituitary hormone influencing reproductive processes is oxytocin. Although well documented as a smooth-muscle stimulating hormone, its role in the male remains uncertain. It may be involved in the dramatic contractions at ejaculation, and may also promote background contractions in the testicular capsule and epididymal duct. In the female, oxytocin is of critical importance at both ends of pregnancy. In the absence of stress during mating or insemination, its reflex release stimulates contractions of the female genitalia, especially of the uterus and oviducts, and facilitates transport of spermatozoa to the site of fertilisation. The release of adrenaline in situations of unease or fright prevents such an action of oxytocin and will

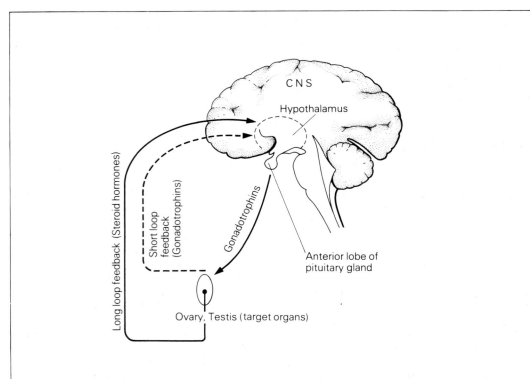

CNS

Hypothalamus

Long loop feedback (Steroid hormones)

Short loop feedback (Gonadotrophins)

Gonadotrophins

Anterior lobe of pituitary gland

Ovary, Testis (target organs)

Figure 1.4 Regulation of hypothalamic and anterior pituitary function in terms of secretion of trophic hormones is largely by feedback actions from target organs located elsewhere in the body. In the case of the reproductive system, gonadal steroid hormones provide the principal form of feedback control although secretion of anterior pituitary hormones themselves is also monitored through short feedback loops

therefore tend to reduce conception rates. Shortly before birth, when the foetal head engages the cervical canal, oxytocin is again released as a reflex and causes the massive contractions of the uterus characteristic of labour. A third neuro-endocrine reflex involving oxytocin occurs at the time of suckling or during milking. In this situation, the hormone acts on the myo-epithelial cells of the mammary gland to cause milk expulsion from the alveoli and thus milk let-down. As the posterior pituitary has nerve connections, neural stimuli leading directly to the release of oxytocin from the nerve terminals can be readily appreciated.

Formation of ovaries, oocyte populations and follicles

The genetic sex of the embryo is fixed at the time of fertilisation, and in mammals the female has an XX sex chromosome complement in contrast to the XY of males. Eggs carry an X sex chromosome, whereas spermatozoa can be either X or Y; the genetic sex of the embryo is therefore determined by the sex chromosome introduced by the fertilising sperm. X- and Y-bearing spermatozoa are produced in equal numbers, and both types are equally competent to fertilise an egg. Accordingly, there should be an equivalent chance of forming male or female embryos at fertilisation, and the sex ratio at this time is termed the *primary sex ratio*.

Completion of fertilisation corresponds to restoration of the diploid condition with appropriate chromosome complements (Table 1.2).

Table 1.2 The diploid ($2n$) and haploid (n) chromosome numbers in farm animals. The diploid complement of chromosomes includes two sex chromosomes

Species	Chromosome	complement
	$2n$	n
Cow	60	30
Sheep	54	27
Pig	38	19

During the first two or three weeks after fertilisation in farm animals, ridges of tissue can be distinguished in the differentiating embryo close to the primitive kidneys, and germ and epithelial cells migrate into these genital ridges to form the gonads. The germ cells therefore have an extra-gonadal origin, and migrate from the yolk sac to the region of the hind gut (Fig. 1.5). In the female, the gonads differentiate into a pair of ovaries with the germ cells taking up an outer (cortical) position; in a male, by contrast, they colonise the medulla and form seminiferous tubules (Fig. 1.6).

The stock of germ cells or oogonia multiplies massively during embryonic and foetal life, such that peak numbers are achieved before the time of birth. The oogonia then give rise to the eggs by commencing a reduction division and, again in

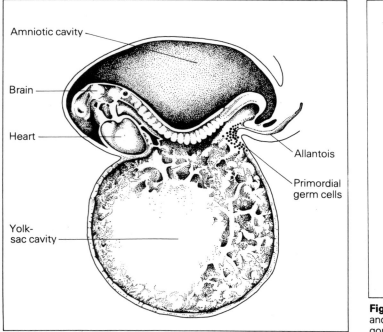

Figure 1.5 The primordial germ cells that provide the vital component of the gonads (i.e. cells that ultimately give rise to eggs or sperm) do not arise in the embryonic gonads but migrate there from elsewhere in the developing embryo, particularly from the membrane termed the yolk sac

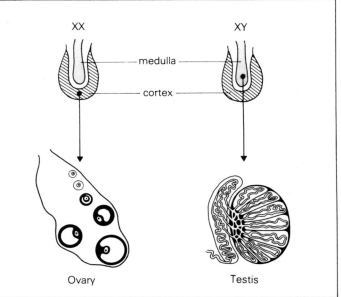

Figure 1.6 The embryonic gonads have an outer portion, the cortex, and an inner portion termed the medulla. Germ cells migrating into the gonads become concentrated in one or other of these regions according to the genetic sex of the embryo. In mammalian females (XX chromosome complement), germ cells aggregate in the cortex and the gonad forms an ovary. In mammalian males (XY chromosome complement), germ cells colonize the medulla and a testis is formed

contrast to the situation in the male, the total number of eggs or oocytes is formed by the time of birth. Thereafter, a progressive wastage or atresia of oocytes is found and, as inferred above, this loss probably begins in late foetal life. Its function is uncertain, but it may represent a form of selection against defective cells. In any event, each ovary of the newborn possesses many thousands of oocytes, of which only a tiny proportion is destined to be shed at the time of ovulation.

After multiplication of oogonia and formation of oocytes (Fig. 1.7), the latter must become surrounded by follicular cells to permit maturation and eventual ovulation. When surrounded by a single layer of cells, this is termed a 'primary follicle', in the multi-layered condition it is a 'secondary follicle', and when a fluid-filled cavity appears this is a 'tertiary follicle' (Fig. 1.8). Oocytes have the potential to be released at ovulation only when within a tertiary (Graafian) follicle, and development of such follicles largely accounts for prepuberal growth of the ovaries. While follicles can thus be regarded as mechanical structures permitting oocyte growth and ovulation, they also have an endocrine function and are largely responsible for the secretion of oestrogen which causes growth of the reproductive tract (see below) and influences the animal's sexual behaviour.

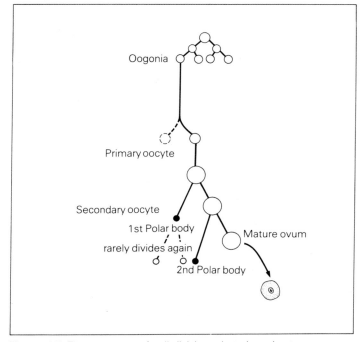

Figure 1.7 The sequence of cell divisions that gives rise to an egg or ovum. Oogonia are diploid cells in the ovary, and the chromosome complement is halved during the divisions of primary and secondary oocytes

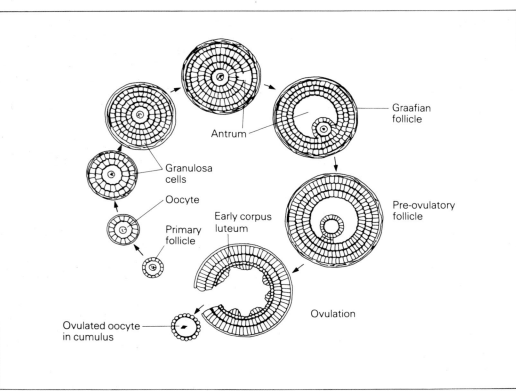

Figure 1.8 Growth of a primary follicle (i.e. an egg surrounded by ovarian cells) by addition and division of somatic cells into a tertiary or Graafian follicle with a fluid-filled antrum. When fully mature (i.e. pre-ovulatory), Graafian follicles respond to pituitary hormone stimulation by collapsing and releasing the egg into the oviduct shortly before fertilisation

Formation and types of genital tract

The very young embryo has the potential for sexual differentiation either as a male or female, having two sets of primitive genital tracts: the Wolffian and Mullerian ducts (Fig. 1.9). Development of the Wolffian ducts would give rise to a male reproductive system, that of the Mullerian ducts to a female system. In normal circumstances and under the influence of the sex chromosome constitution of the embryo, one set of ducts develops while the other becomes vestigial. However, the Mullerian ducts develop not because the embryo possesses two X chromosomes, but rather as a permissive state in the absence of the Y chromosome. Maleness is the dominant condition, and in the absence of a Y chromosome, and more specifically in the absence of male sex hormones (androgens) from the embryonic testes, the Wolffian ducts regress and the Mullerian ducts become prominent.

Growth of these ducts proceeds, and before the time of birth the component parts of the female genital tract can be distinguished as four regions. These are the paired oviducts, the uterine horns and body, the cervix and the vagina. Within a species, each of these regions has a characteristic anatomy and histology, and each performs specialised physiological functions. Varying degrees of fusion are found in the uterus of farm animals, and a reduction in the extent of the uterine horns, together with an increase in size of the uterine body, is associated with a reduction in the number of potential offspring. Thus, a pig has two long uterine horns and a negligible uterine body whereas a horse has these proportions essentially reversed (Fig. 1.10); the situation found in cattle and sheep is intermediate, with a fold or septum extending into the uterine body. Accessory glands are also associated with the reproductive tract – especially in the walls of the cervix and vagina – but these remain small when compared with development of accessory glands in the male (see Ch. 2, p. 22). Even so, the cervical secretion of mucus is important at mating, during pregnancy, and for the events of birth.

Puberty

As noted above, ovarian follicles secrete oestrogens before puberty and these contribute to growth of the genitalia. However, there is a marked increase in follicular activity shortly before puberty, this being a reflection of increasing gonadotrophin secretion. Although changes in the amount and manner of pituitary gonadotrophin secretion at puberty can be described with accuracy, the underlying causes remain to be specified. Perhaps the most favoured explanation is that during the interval from birth to puberty, the central nervous system inhibits those regions of the hypothalamus concerned with gonadotrophin releasing hormone(s), and that this inhibition is removed when the animal has achieved sufficient

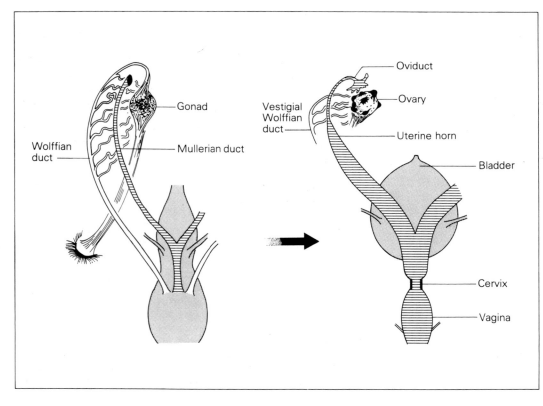

Figure 1.9 Generalized diagram to show formation of the female reproductive system. The very young embryo has the potential to form either a male or a female genital system since it has a rudimentary development of both male (Wolffian) and female (Mullerian) duct systems. The Wolffian ducts regress in female embryos whilst the Mullerian ducts give rise to paired oviducts and uterine horns, a cervix and a vagina

female is seen as sexual receptivity to mounting and mating, and is best defined as the age at which oestrus (i.e. heat) is first detected and is followed by characteristic cyclic activity in the non-pregnant animal. In the absence of a mature male, oestrus may be detected by sexual excitement and mounting or riding between females, together with a swelling of the external genitalia and the secretion of cervical mucus seen on the vulva. The approximate age of animals at puberty is indicated in Table 1.3. Variation is attributable to genetic differences such as those of breed or crossbreeding: smaller breeds reach

Table 1.3 Some commonly recorded figures for the age at puberty. The figures are very much influenced by breed or strain and by environmental conditions – especially level of nutrition

| Species | Age at puberty (months) | |
	Range	Mean
Cow	8.0–17	10.5
Sheep	4.5–15*	7.5
Pig	5.0–8	7.0

*Depends on season of birth. Early born ewe-lambs will reach puberty the same year; late-born lambs may not commence oestrous cycles until the autumn of the following year.

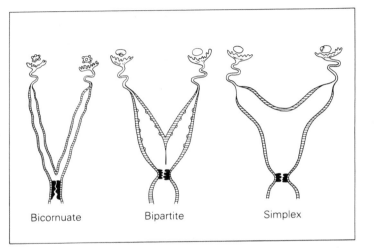

Figure 1.10 The overall form of the female reproductive tract varies according to species, with two long horns in the pig (bicornuate uterus), shorter uterine horns and a larger uterine body in cows and sheep (bipartite uterus), and a large uterine body in the mare and primates. This progressive modification of the uterus matches a reduction in the potential number of offspring

Bicornuate   Bipartite   Simplex

body maturity. This stage can usually be expressed as a proportion of the animal's mature weight, and is represented by the point of inflexion on the growth curve.

Turning to events of more direct interest, puberty in the

puberty before the larger, slower maturing breeds while inbreeding tends to delay and crossbreeding to accelerate the onset of puberty. Feeding regime and, thereby, nutritional status, season of the year, housing or stocking density, and the presence or absence of a mature male all influence the onset of puberty. Puberty is usually brought forward in groups of females exposed to an experienced male, and generally occurs earlier in the spring and summer than autumn or winter.

Oestrous cycle, behaviour and detection

In females not exposed to a male or artificially inseminated, puberty leads to a repeated cycle of ovarian events that finds expression in the oestrous cycle of farm and laboratory animals or in the menstrual cycle of primates. Oestrous cycles are regulated by pituitary and ovarian hormones, and are of characteristic mean length for a species varying from 16 or 17 days in sheep to 21 days in cattle and pigs (Table 1.4). A

Table 1.4   Typical figures for the duration of oestrous cycles

| Species | Duration of oestrous cycle (days) | |
	Range	Mean
Cow	20–22	21
Sheep	16–17	16.5
Pig	20–22	21

convenient starting point from which to describe the cycle is ovulation. During this event, the egg is released from a Graafian follicle into the oviduct, and a corpus luteum

commences to form within the collapsed follicle (Fig. 1.11). The maturing follicle has been actively secreting a steroid hormone, oestrogen, and it is principally under the influence of oestrogen that the animal comes into heat shortly before ovulation. The function of the corpus luteum is to secrete another steroid hormone – progesterone – and this it does in increasing amounts (Fig. 1.12) as the structure grows into a solid mass of luteal tissue. Under the influence of progesterone, the reproductive tract is prepared for the outcome of a successful mating; that is, nourishment of a developing embryo. This is due to the influence of the steroid on oviducal and, more particularly, uterine secretions, and involves proliferation of the uterine glands.

Progesterone is secreted for at least two-thirds of the oestrous cycle, this being referred to as the *luteal phase*. If the animal has not mated successfully and embryos are not present in the uterus – and the animal prolongs monitoring this situation as late as possible within the characteristic cycle length – then the corpus luteum is caused to regress rapidly by uterine prostaglandin secretion. The specific hormone is prostaglandin $F_{2\alpha}$ ($PGF_{2\alpha}$), and progesterone synthesis is curtailed by this luteolytic activity. One or more Graafian follicles is then stimulated to mature under the influence of pituitary gonadotrophin secretion, follicular oestrogens are secreted in increasing concentrations, and the animal returns to the receptive condition of oestrus or heat. This *follicular phase* of the cycle is of short duration, the underlying strategy

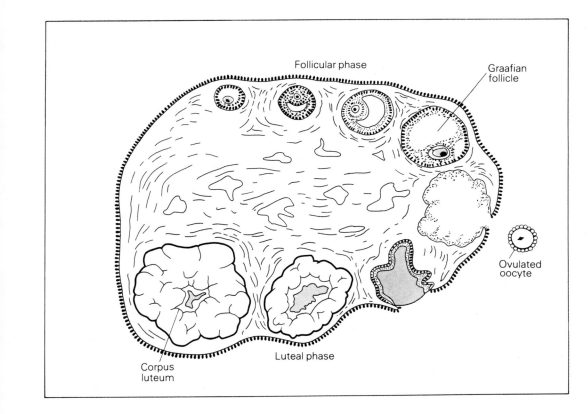

Follicular phase

Graafian follicle

Ovulated oocyte

Corpus luteum

Luteal phase

Figure 1.11 Simplified section through a mammalian ovary to illustrate the preponderance of follicular growth and maturation during the follicular phase and, after ovulation, the development of the corpus luteum during the relatively longer luteal phase

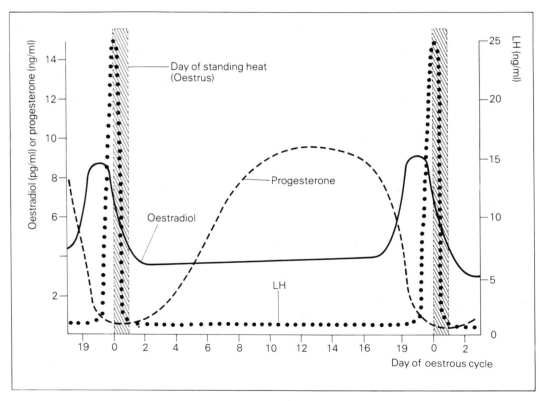

Figure 1.12 Simplified diagram of the changing concentrations of oestrogen (oestradiol) and progesterone during the 21-day oestrous cycle of cattle, and also showing the pre-ovulatory surge of LH that occurs very close to the onset of oestrus

being to return the female to mating condition as rapidly as possible after failing to detect a viable embryo in the uterus. Active follicular growth is prevented during the luteal phase by a negative feedback influence of progesterone upon pituitary gonadotrophin secretion, but as soon as circulating progesterone levels begin to fall, stimulation of terminal growth in one or more follicles can resume. Even so, there are periods of gross ovarian inactivity in some species, the period of seasonal quiescence in sheep and certain other species being referred to as anoestrus. Cattle and pigs, by contrast, are usually polyoestrous and show cycles throughout the year.

The ovarian and uterine events underlying the oestrous cycle are of little direct interest to the producer, nor are the related changes in the histology of the tract, especially in the vaginal epithelium. What is important in practice is to identify the correct time for mating or insemination, and it is the interplay of oestrogen and progesterone that determines the animal's sexual behaviour. Progesterone dominance causes the animal to be uninterested in, and refractory to, the male and this is the situation for the major part of the cycle – the dioestrous interval. However, increasing sexual activity and excitement are seen under the influence of oestrogen as the follicle(s) matures, and this phase of a few days during which vulval swelling becomes conspicuous is termed 'pro-oestrus' Oestrus is due to an enhanced influence of oestrogen on the brain and hypothalamus, although these structures must first have been primed with progesterone. Willingness of the

Standing to back pressure in boar's presence

Figure 1.13 Among the farm species, sexually receptive females are identified most easily in pigs: those that are in oestrus will respond to pressure on the back, especially in the presence of a mature male, by immobility, arching of the back, and 'pricking' of the ears. This test is not applicable in cows and sheep – in which species oestrus is best identified with a teaser male

female to stand for mounting and mating is the only true sign of oestrus. Among the farm species, oestrus can be demonstrated most clearly in the pig with rigidity of the back, pricking of the ears (Fig. 1.13) and a characteristic 'growling' sound. In the absence of a male, animals in oestrus will usually accept mounting by other females. To put these events in perspective, oestrus occurs when the Graafian follicle is most mature and ovulation is imminent. In this way, mating and deposition of sperm in the female tract are closely synchronised with release of the egg(s) from the ovary.

Timing of ovulation

As already noted, ovulation is release of the egg from a Graafian follicle during collapse of this ovarian structure. The word collapse is used advisedly since, in farm animals at least, mature follicles do not rupture but rather undergo a reduction in intra-follicular pressure and become soft and flaccid in the hour or two preceding ovulation. Enzymatic changes in the wall of the follicle undoubtedly underlie this condition, and much attention has focused on the role of collagenase in causing a dissociation and eventual breakdown of cell layers near the apex of the follicle. There is good evidence also for smooth muscle activity in the wall of the follicle, probably to ensure expulsion of the egg when it has not been fully displaced from the antrum (Fig. 1.8, p. 10) during the outflow of follicular fluid.

Although ovulation is spontaneous in farm animals in the sense that it does not require to be induced by a coital stimulus – as is the case, for example, in rabbits, cats and ferrets – it does need to be triggered by a specific and discrete hormonal stimulus from the anterior pituitary gland. This is in the form of a surge or spike-release of the gonadotrophins, FSH and LH, with the magnitude of the LH release being much greater. The gonadotrophin binds to the wall of the follicle and provokes the sequence of events leading to follicular collapse at a specific interval after the stimulus (Table 1.5). The surge

Table 1.5   The timing of ovulation in relation to the duration of oestrus and to the pre-ovulatory surge of gonadotrophic hormones that occurs close to the onset of oestrus

Species	Duration of oestrus (hours)	Timing of ovulation (hours)	Timing of ovulation after gonadotrophin surge (hours)
Cow	12–26	10–12 after *end* of oestrus	30–36
Sheep	24–36	24–26 after *onset* of oestrus	25–26
Pig	30–>60	36–40 after *onset* of oestrus	40–42

of pituitary gonadotrophin secretion has itself been brought about due to a positive feedback influence of follicular oestrogens which, as pointed out, is also responsible for stimulating the onset of oestrous behaviour. In a sense,

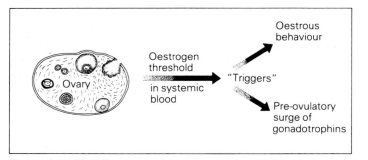

Figure 1.14 To indicate two major influences on the hypothalamus of the high concentration of oestrogens secreted by mature follicles. Oestrogens act on the hypothalamus to provoke oestrous behaviour and they also initiate the surge of gonadotrophic hormones (LH and FSH) at the onset of oestrus that causes ovulation. Receptivity to the male for mating and release of the egg(s) are thereby functionally related

therefore, it is follicle maturity, expressed as a threshold of oestrogen secretion, that regulates both oestrous behaviour and the time of ovulation (Fig. 1.14). The important point in practice is that the surge of gonadotrophin is found at or closely coincident with the onset of oestrus, and ovulation occurs at a known interval thereafter (Table 1.5). This interval is 25–26 hours in sheep, 30–36 hours in cattle and 40–42 hours in pigs.

If the onset of oestrus could be accurately determined, it follows that the time of ovulation could be predicted. However, recognising the onset of oestrus requires unstressed animals to be tested with an experienced teaser male every two or three hours until they stand for mating, and clearly this would not be feasible in farming practice. Therefore, when overt oestrus is identified on the farm, 6–8 hours or more have probably elapsed since its actual onset and a similar interval after the gonadotrophin surge, so the animal is that much closer to the time of ovulation. In fact, because the duration of oestrus or heat in sheep closely parallels the interval from the LH surge to ovulation, release of the egg is found just at or shortly after the end of heat. In cattle, ovulation occurs 10–12 hours after the end of heat whereas in pigs, by contrast, ovulation occurs on the second day of heat, possibly 12 or more hours before it ends. In situations of 'hand mating', as well as in artificial insemination, there is thus a considerable danger of mating pigs *after* ovulation has occurred: the deleterious effects of egg ageing may already have commenced and a smaller litter size will result (see Ch. 3, p. 53).

As to the number of follicles ovulating, this is clearly influenced genetically as seen in the differences between species, breed and even strain. Cattle generally bring a single follicle to maturity at each oestrus and therefore release a single egg at ovulation, sheep mature 1–3 follicles and pigs a number between 11 and 24. In pigs, in particular, age after puberty also has an influence on this figure, or *ovulation rate* as it is loosely termed, with the number of eggs shed increasing by approximately one per cycle for the first seven or eight cycles. This fact puts the stockman in the dilemma of having to choose whether to let gilts undergo a series of barren cycles in order to increase ovulation rate, while bearing the cost of maintenance over each of these periods of three weeks, or whether to breed soon after puberty and settle for a smaller potential litter size. Good nutritional status also tends to increase ovulation rate, and is the reason underlying the practice of flushing shortly before breeding time in sheep.

The male, semen production and artificial insemination

2

Vigorous, mature males have important social roles in the life history of mammals; but while a mature male may be valuable in stimulating puberty and the onset of oestrous cycles in farm animals, his one essential function is to mate receptive females so that fertilisation and embryonic development may follow. On many livestock units, stud males are no longer kept since their function has been taken over by artificial insemination. In this technique, appropriate samples of fresh or stored semen are introduced into the female tract by a simple manual procedure. However, before considering the sequence of events at mating or insemination, an understanding of the formation and function of the male reproductive system is desirable. We must examine how the maleness of an offspring is established, and the changes that lead to semen production.

Development of male reproductive system

Genetic sex

As already described in Chapter 1, the genetic sex of the future offspring is determined at the time of fertilisation. In the case of a male embryo, the egg which carries an X sex chromosome would have been penetrated by a Y-bearing sperm to give the XY sex chromosome complement characteristic of male mammals. Since the ejaculate contains equal numbers of X- and Y-bearing spermatozoa, there should be a 50 : 50 chance of producing a male embryo at fertilisation, even though the ratio of males to females at birth may not be identical when examined in an adequately sized population.

Formation of ducts

Under the influence of the XY sex chromosome complement, the embryonic gonads differentiate into primitive testes and these, in turn, regulate development of the male reproductive system (the Wolffian ducts) and regression of the female system (the Mullerian ducts; see Ch. 1, p. 11). The key factor in this instance is the dominating role of the Y chromosome and the secretion of male sex hormone by the embryonic testes; this stimulates development of the male ducts while another hormone from the testes actively suppresses growth and differentiation of the potential female system (Fig. 2.1). The developed male system consists of ducts leading from the testis, an epididymis and vas deferens all as paired structures, a single pelvic urethra from fusion of the two ducts, and a further portion of the urethra within the penis (Fig. 2.2). There is thus a continuous conduit from each testis to the exterior. The functions of the epididymis and vas deferens are concerned with ripening, storage and ejaculation of spermatozoa. In close association with this duct system are the accessory glands that together produce a fluid – the seminal plasma – in which spermatozoa are ejaculated. The principal accessory glands are the paired seminal vesicles, the prostate, and the bulbo-urethral or Cowper's glands (Fig. 2.2); their production of seminal plasma is controlled by testicular androgens.

Formation of gonads

The gonads themselves can be distinguished in the very early embryo as paired ridges of genital tissue into which the large germ cells migrate from elsewhere in the embryo. The cells become the stem cells or spermatogonia and ultimately give rise to the spermatozoa. In the embryo and foetus, however, they simply multiply to become arranged around the inner wall of the seminiferous tubules along with protective nurse cells (Fig. 2.3), and sperm formation does not commence until the time of puberty many months later in sheep and pigs or, indeed, a year or more later in cattle. By contrast, the other major tissue of the developing testes is active early in embryonic life, producing hormones rather than cells, and is the source of the male sex hormones, especially testosterone; it is termed the interstitial tissue and lies between the seminiferous tubules (Fig. 2.3).

Growth

Growth and development of the reproductive system takes place throughout foetal life so that the component parts are distinguishable to the naked eye upon dissection at birth (Fig.

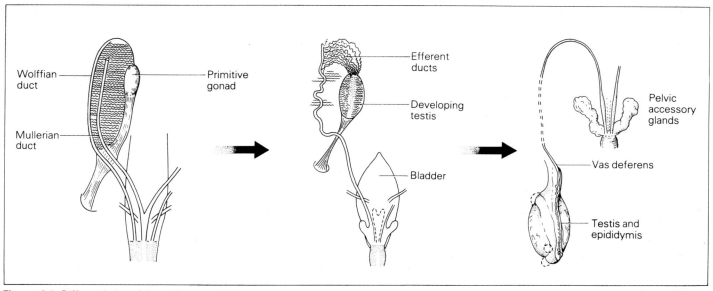

Figure 2.1 Differentiation of the male reproductive system by development of the Wolffian ducts and regression of the Mullerian ducts. The male system consists of paired efferent ducts, epididymis and vas deferens which converge at the urethra in the region of the accessory glands (e.g. prostate and seminal vesicles)

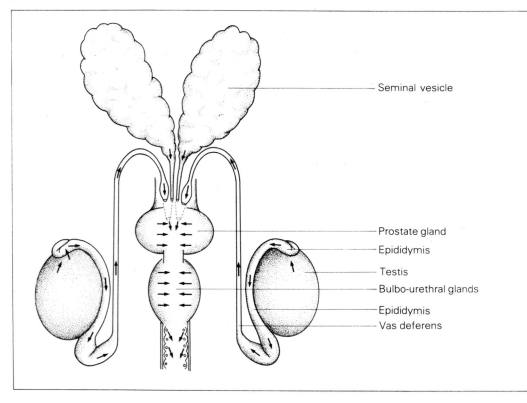

Seminal vesicle

Prostate gland

Epididymis

Testis

Bulbo-urethral glands

Epididymis

Vas deferens

Figure 2.2 Generalised diagram of a fully differentiated male reproductive system, indicating the three principal accessory glands – the seminal vesicles, prostate and bulbo-urethral glands

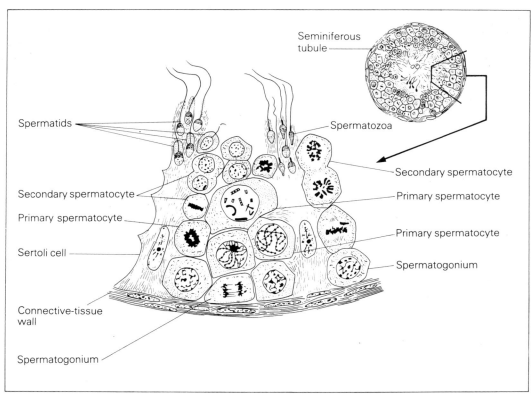

Spermatids

Spermatozoa

Secondary spermatocyte

Secondary spermatocyte

Primary spermatocyte

Primary spermatocyte

Primary spermatocyte

Sertoli cell

Spermatogonium

Connective-tissue wall

Spermatogonium

Seminiferous tubule

Figure 2.3 Sections through testicular tissue to illustrate the process of spermatogenesis and the role of the Sertoli cells in supporting the germ cells. The interstitial tissue lies between individual seminiferous tubules

2.6). The testes have already descended through the inguinal canals from the body cavity into the scrotal sac. Although they have secreted sex hormones since early in embryonic life, it is only shortly before puberty that androgen output from the much-enlarged testes increases dramatically, giving rise to the pronounced secondary sex characteristics of the male. These include changes in body conformation, growth of hair around the preputial orifice, development of a typical male 'voice' and odour, and an increase in mounting and aggressive tendencies.

Events at puberty: how are sperm produced?

Sperm production heralds the onset of puberty. Formation of spermatozoa requires a series of specialised cell divisions within the seminiferous tubules, commencing with spermatogonia and passing through the stages of spermatocytes and spermatids (Fig. 2.3). This sequence of cell types is supported and nourished by contact with nurse cells of the tubules, the Sertoli cells. These changes give rise to a highly specialised cell, the spermatozoon (Fig. 2.4), with half the number of chromosomes for the species, only a droplet of residual cytoplasm, and a distinct head, mid-piece and tail that in due course, enable the cell to swim. During the process of spermatogenesis, therefore, a highly modified cell is produced whose eventual purpose is to deliver its nucleus into the egg at fertilisation. Because the parent cells contain X and Y sex

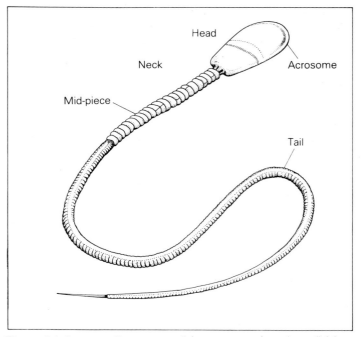

Figure 2.4 A mammalian sperm cell (spermatozoon) to show division into head, mid-piece and tail. The head contains a rigid nucleus and acrosomal enzymes to facilitate penetration of the egg membranes. Energy is generated in the spirally-arranged mid-piece, whilst propulsive force is derived from the whip-lash movements of the tail

chromosomes, these divisions lead to equal numbers of X- and Y-bearing spermatozoa; they are liberated into the lumen of the seminiferous tubules.

The changes that produce spermatozoa take place in the testes from where the cells are carried by fluid flow as a dilute suspension into the adjoining epididymis. This is an extremely long (>100 feet in a bull and boar) highly convoluted tube arranged on the surface of the testis (Fig. 2.5) in which the sperm suspension is concentrated into a creamy mass and there then follows a process of ripening and storage. As can be demonstrated experimentally, spermatozoa entering the epididymis from the testis are quite unable to fertilise eggs or even to show swimming ability, whereas after 11–18 days of passage along the epididymal duct, they have matured and acquired both these characteristics. Perhaps it should be emphasised that they have acquired swimming and fertilising *potential*, for in fact spermatozoa remain completely immotile in the epididymis and first exhibit motility at the time of ejaculation. Other maturational changes include loss of the droplet of residual spermatid cytoplasm and development of a distinct distribution of charges on the sperm surface. The head becomes more rigid and the acrosome is more intimately applied to the nucleus.

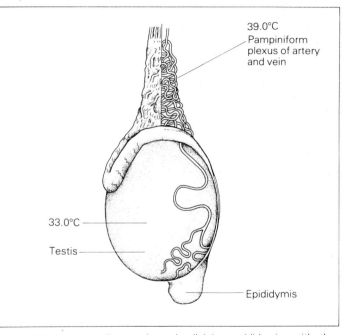

39.0°C
Pampiniform plexus of artery and vein

33.0°C

Testis

Epididymis

Figure 2.5 A mammalian testis and adjoining epididymis, with the suspensory portion dissected to show the arrangement of the pampiniform plexus. The testicular artery and vein are intimately interwoven in this region, and there is a counter-current exchange of heat so that arterial blood enters the testis 4-7°C below abdominal temperature

Temperature control: why scrotal testes?

Normal production and ripening of spermatozoa, as described in the above sequence of events, can only take place at a temperature several degrees below that of the abdomen. This explains the requirement for the testes and adjoining epididymides to be suspended in a scrotal sac. If the testes fail to descend into the scrotum, the process of spermatogenesis likewise fails and the male is infertile – even though he may mount and produce seminal fluids at ejaculation. This condition of abdominal testes (cryptorchidism) is quite common in horses and boars, but much less so in other farm mammals. Infertility is not usually caused if only one of the testes is concerned: the outcome here is a lower concentration of spermatozoa in the ejaculate.

There is a reduction of 4–7 °C between abdominal and testicular temperatures (see Fig. 2.5), regulation of the difference being obtained primarily by the cremaster muscles that support the testes. The warmer the ambient temperature, the more relaxed the cremaster muscles giving the testes and scrotum a pendulous condition, thereby increasing their distance from the body cavity and cooling the blood within. An added sophistication is that blood returning to the abdomen in the testicular venous network functions to cool the incoming arterial blood. This is achieved by means of extreme coiling of the vein around the testicular artery (see Fig. 2.5). These various specialisations stress the requirement of a reduced scrotal temperature for normal sperm production in farm animals, and they endorse the value of removing undue wool growth from the scrotal sac of rams.

Sperm production and loss: is production continuous?

Excluding instances of disease and nutritional deficiencies, production of spermatozoa continues from puberty to old age. The only exception to this is in animals that are seasonal breeders, such as deer and to a lesser extent sheep, in which the cell divisions in the seminiferous tubules may be diminished or arrested. There is thus a continuous passage of spermatozoa from the testis into the epididymis, a phase of ripening during passage along the epididymal duct, and then liberation through leakage or ejaculation. Spermatozoa are stored in the lowermost segment of the epididymis and in the vas deferens prior to ejaculation; undue congestion of these ducts is avoided by progressive leakage of spermatozoa into the bladder from where they are voided in the urine. In the absence of sufficient mating activity, masturbation is another means of reducing sperm reserves and occurs in all species of farm animal.

Numerous estimates of the rate of sperm production are recorded in the literature, one of the most impressive statistics being that one gram of testicular tissue in a mature bull may produce nine million spermatozoa per 24 hours. However, the testes are not autonomous, and ultimate control of their activity resides in hormones of the anterior pituitary gland – especially in secretion of FSH, LH and prolactin. There may also be feedback of a non-steroidal compound from the testes (inhibin) which regulates secretion of the gonadotrophins and prolactin and, therefore, to some extent the rate of sperm production.

Androgen production by the interstitial tissue seems to be regulated mainly by LH secretion, but males do not show cyclic peaks of gonadotrophin secretion comparable to the pre-ovulatory surges of females. The ability to secrete gonadotrophins in this manner has been abolished by an influence of testicular hormones on the brain and hypothalamus during a critical period of development.

Events at ejaculation

The phenomenon of ejaculation has many components, although these can certainly be reduced in number and simplified during semen collection procedures for artificial insemination. The principal physical event in the genital tract at ejaculation is an almost instantaneous mixing in the region of the pelvic urethra of a dense mass of spermatozoa with the seminal plasma contents of the accessory glands to give the fluid expelled from the penis, the semen. Involuntary smooth muscle contractions propel spermatozoa from the tail of the epididymis and vas deferens to the urethra, and it is the powerful contractions of the latter that deliver the semen at

ejaculation. The temperature of semen corresponds closely with that of the abdomen, and exposure to a raised temperature is one of the factors stimulating motility in liberated spermatozoa. Also important are the sugar (fructose) and other metabolic substrates in seminal plasma, together with the reduced density of the sperm suspension when mixed with the accessory secretions.

At the time of ejaculation, the form of the extended penis can be fully distinguished. In contrast to the 'spongy' penis of

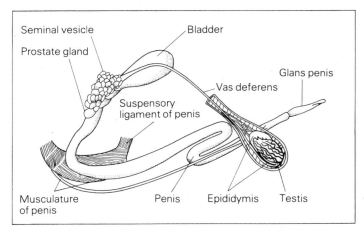

Figure 2.6 The reproductive system of a bull to show the sigmoid flexure of the penis that permits extrusion at the time of mating

the stallion which becomes engorged with blood and enlarges prominently during erection, the penis of the bull, ram and boar is fibro-elastic. It changes little in *actual* size during mating, its *apparent* increase in size being due to relaxation of the muscle regulating the sigmoid flexure (Fig. 2.6); this permits extrusion and straightening of the penis for introduction into the female tract.

Semen characteristics

Some principal characteristics of the ejaculate for the major farm species are listed in Table 2.1. Across species, it will be

Table 2.1 Characteristics of semen ejaculated by mature males together with recommended frequencies of collection. Note particularly the volume of semen and numbers of spermatozoa emitted by the boar

Species	Volume of ejaculate (ml)	Sperm concentration ($\times 10^6$/ml)	Total sperm per ejaculation ($\times 10^9$)	No. of collections per week
Bull	4–8	1 200–1 800	4–14	3–4
Ram	0.8–1.2	2 000–3 000	2–4	12–20
Boar	150–500*	250–350	40–50	2–3

*Includes gelatinous secretion of the accessory glands.

noted that there is no obvious correlation between body size (i.e. liveweight) and volume of the ejaculate, for the bull produces 4–8 ml of semen while the boar may give 200–400 ml. Within a species, however, increase in body size is generally expressed as an increase in ejaculate volume, doubtless due to growth of the accessory glands. Other factors influencing the volume of the ejaculate are the frequency of mating or semen collection and the extent of foreplay or 'teasing'. Whereas an increased duration of teasing generally results in a larger semen volume, a greater frequency of mating tends to reduce the volume of the ejaculate and increase the proportion of immature spermatozoa. Nonetheless, it should be possible to use a mature boar 2–3 times per week and a bull 3–4 times per week provided the animals are maintained under good conditions of husbandry.

While the foregoing remarks have concentrated on semen volume, the vital component of the ejaculate is, of course, the spermatozoa. Table 2.1 shows that the total number of spermatozoa in the ejaculate varies much less than the overall volume, even though the boar ejaculate still contains more spermatozoa than the bull ejaculate. Total sperm production in a species increases with body growth, reflecting increased testicular size and an increase in the seminiferous epithelium.

Artificial methods of semen collection

A variety of methods for semen collection from farm animals has been described during the last 60 years, and these involve direct stimulation of the musculature of the male tract or simulating the influence of the vagina on the glans penis. The latter approach requires the male to mount a teaser female or a dummy. The technique in the former is usually electro-ejaculation and may be especially suitable for animals no longer able to mount due, for example, to damaged feet and yet whose semen is known to be valuable. A bipolar probe is introduced into the rectum and pulsations of low voltage stimulate the musculature of the reproductive tract and in due course ejaculation, even though the penis will not be fully extended. While animals vary in their sensitivity to this approach, it is successful in bulls and rams but much less so in boars or stallions. There is no deleterious influence of the technique on the spermatozoa, but the semen produced is more dilute than after mounting or mating.

The alternative approach – used when the male mounts a teaser or dummy – involves an artificial vagina or, less commonly, direct grasping of the penis. The artificial vagina is a rigid cylinder with a thin rubber liner separated from its wall by a fluid reservoir (Fig. 2.7). This is filled with water at 40–42 °C until the lining is appropriately stretched, so that the temperature and pressure of the device closely mimic the influence of the vagina. An insulated and calibrated glass vessel is attached to one end, into which the semen passes during and after ejaculation. Use of an artificial vagina is widespread for semen collection from bulls and stallions, but somewhat less so for rams which respond well to

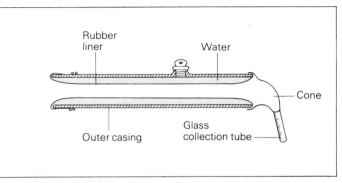

Figure 2.7 An artificial vagina used for semen collection from bulls. Water, slightly above body temperature, separates a soft rubber liner from the outer casing of the device. Temperature and pressure requirements vary slightly between bulls.

electro-ejaculation. Extensive observation at artificial insemination stations indicates that males have strong individual preferences for the routine that immediately precedes semen collection and indeed for the condition of the artificial vagina itself: the degree of tension in the rubber lining and the presence or absence of lubrication are two such examples.

A teaser female is either a nymphomaniac cow or one maintained permanently in heat by means of injection of an oestrogen solution in oil. A dummy, on the other hand, need bear no close anatomical resemblance to the species in question but is a well-padded structure of appropriate dimensions arranged on three or four steel legs (Fig. 2.8). An essential feature of the dummy is that it should be well impregnated with odoriferous secretions, such as preputial fluid, saliva and stale urine. Paradoxically, secretions from other males may be found more stimulating than cervical or vaginal mucus from oestrous females, though the routine followed by the male prior to inspecting and mounting the teaser female or dummy may be at least as important as olfactory stimulation. When the male mounts and commences to thrust, the penis is deflected into the artificial vagina held alongside the flank and the ejaculate collected by immediate tilting of the device after service towards the glass receptacle. In the case of boars, an artificial vagina is not widely used these days since a simpler approach is available and also because stale urine and bacteria from the preputial sac tend to run down the surface of the penis and contaminate the semen during the prolonged period of ejaculation. Instead, the spirally arranged glans penis is grasped manually during the boar's initial thrusting movements, and pressure applied to the tip should cause full extrusion of the penis and ejaculation. Unlike the situation in bulls, rams and stallions, temperature is not critical to the boar's ejaculatory reflex.

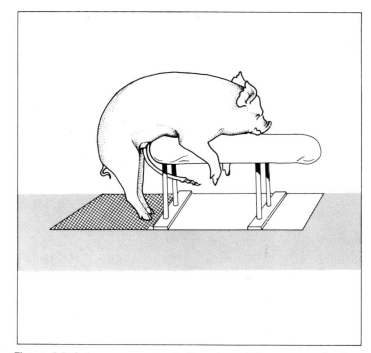

Figure 2.8 A dummy sow mounted by a boar during semen collection procedures for artificial insemination. Once trained to mount such a dummy, there is little problem in obtaining semen collections on a routine basis

Examination of semen samples

As inferred above, it is extremely important to avoid chilling or any other form of temperature shock to the semen during collection and subsequent handling. Spermatozoa are notoriously sensitive to sudden changes of temperature *in vitro*, and although they can be cooled and indeed deep-frozen, such procedures involve controlled and progressive changes of temperature – usually of diluted semen and invariably in the presence of protective agents. Accordingly, after collection in an insulated glass vessel, the ejaculate is transferred to a processing laboratory where it is examined under appropriate conditions. Superficial examination of the semen sample will reveal whether there is contamination with dirt, hair, blood or urine. Blood may indicate some damage to the penis during collection, whereas urine suggests an unsuitable temperature in the artificial vagina. In the case of boar semen, the prominent gelatinous fraction is immediately separated off by filtration of the ejaculate through muslin or cotton gauze; otherwise the gel would absorb fluid, rapidly reducing the volume of semen and, to a lesser extent, the number of spermatozoa. This procedure is essential even if, during the prolonged period of ejaculation (5–15 min), the semen has been collected in its various fractions of an initial watery secretion, then a sperm-rich, followed once more by an extremely dilute sperm suspension (Fig. 2.9). The gelatinous fraction is also filtered from stallion semen.

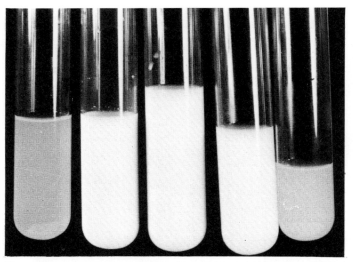

Figure 2.9 A fractionated ejaculate from the boar to show the differing concentrations of sperm in the pre-sperm, sperm-rich and post-sperm fractions. Dilution of the sperm suspension depends on the relative contribution of fluids from the accessory glands

Microscopic examination of the semen sample includes an estimate of the proportion of actively motile spermatozoa and measurement of the concentration of spermatozoa. Motility is examined in a droplet of semen smeared on a heated microscope slide with the microscope in a perspex incubator at

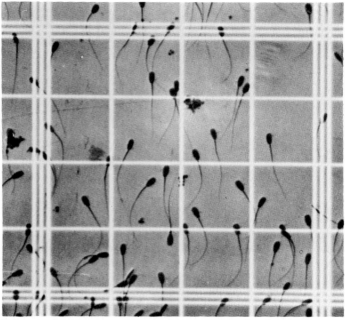

Figure 2.10 A haemocytometer slide used for estimating the concentration of spermatozoa in a semen sample, and thereby its dilution potential for use in artificial insemination. Each square is of known area, and the depth between coverslip and slide is also known, enabling cells per unit volume to be calculated

37 °C, or at least having a heated stage. Good samples of ram and bull semen show a swirling, wave-like motion due to the very high concentration of active cells. The concentration of spermatozoa is measured either in a ruled counting chamber on a special microscope slide (a haemocytometer slide: Fig. 2.10) or, more simply and rapidly, in a calibrated optical instrument used routinely at artificial insemination stations. In this instrument, the density of the sperm suspension is related to the interference with transmission of a beam of light. The reason for measuring the density of spermatozoa, and thus the total number in a known volume of ejaculate, is that this directly influences the extent to which the semen can be diluted and, therefore, the number of insemination doses available. During all these procedures it is important that samples of semen are not exposed to direct sunlight or to fumes from volatile chemicals or disinfectants.

Using a diluted sample, the morphology of the spermatozoa is also examined so that the sample can be rejected if there are many cells of unusual shape, with damaged heads or tails or, exceptionally, when these components are separated (Fig. 2.11). In addition to the criterion of motility, the proportion of live to dead cells can be checked by a staining method in which spermatozoa with a damaged membrane (i.e. dead or moribund) are permeable to the stain. Calculation of this proportion again influences the dilution potential of the ejaculate.

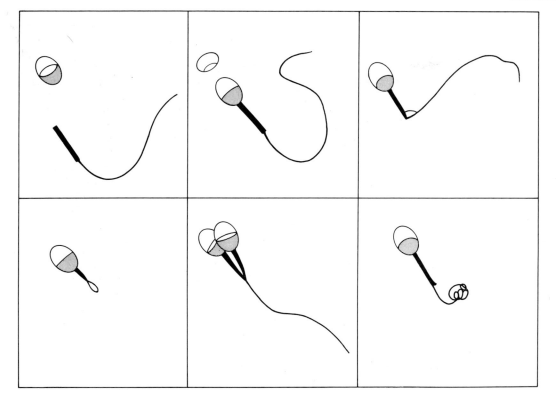

Figure 2.11 Abnormal forms of sperm morphology observed in ejaculates from farm animals. Abnormalities of the head, midpiece or tail are invariably associated with diminished fertility or even sterility

Dilution of semen

Dilution of semen has, as its overall objective, the more widespread use of a male's fertilising ability than would be possible by natural service (see Table 2.2); this is stressed in

Table 2.2 To illustrate the potential for dilution and insemination of single ejaculates from mature animals

Species	No. of inseminations per ejaculate	Volume of inseminate after dilution (ml)	No. of motile sperm ($\times 10^6$)
Bull	400	0.25–1.0	5–15
Ram	40–60	0.05–0.2	50
Boar	15–30	50–100	2 000

the North American word 'extender' for semen diluents, and relies on the fact that a single ejaculate contains many times more spermatozoa than are required for fertilisation. The essential feature of a diluent is that it should have a negligible deleterious influence on the motility, fertilising ability and, indeed, overall viability of individual spermatozoa. Commonly, a diluent also includes one or more protective agents so that the step of diluting the ejaculate can also serve as a preliminary to preservation. The constituents of a diluent would include the following:

1. a metabolic or nutritional substrate – usually a sugar;

2. salts in suitable concentrations to buffer the sperm suspension against changes in pH and osmotic pressure;
3. a large molecular weight component to protect the sperm cells against the damaging effects of chilling; and
4. antibiotics to prevent the growth of bacteria.

Examples of components in a medium that has been widely used for many years are glucose, citrate and bicarbonate, egg yolk and antibiotics, and the medium is saturated with gaseous carbon dioxide thereby immobilising the spermatozoa. After dilution with media containing such essential components, bull semen may then be cooled to temperatures close to but above 0 °C and boar semen to 5 °C. The decrease in temperature reduces the motility and associated energy loss of the sperm cells, enabling chilled bull or boar semen to be used satisfactorily in artificial insemination for at least three days – even though there is some decline in fertilising capacity.

Deep-frozen storage

The dominant principle during freezing and thawing spermatozoa is to avoid damaging the cell during the separate steps of cooling and warming. It is achieved by means of a protective agent in the dilution medium which penetrates the sperm membranes during a period of equilibration; this serves to prevent the damaging effects of intracellular formation of ice. Glycerol has been the traditional protective agent since

the first successful freezing experiments with spermatozoa in the late 1940s, but other protective agents, such as dimethyl sulphoxide, have also found favour. Important factors during the overall procedure of deep-freeze preservation include:
1. concentration of protective agent in the diluent;
2. time and temperature of equilibration with the protective agent;
3. rate of cooling and thawing of the sample;
4. means of removal of the protective agent (e.g. dilution).

Frozen semen was previously stored at the temperature of solid carbon dioxide ($-79\,°C$), but liquid nitrogen ($-196\,°C$) is now used extensively with the semen samples held in suitable containers either actually in the liquid or in its vapour (Fig. 2.12). At most, there is a very slight and slow deterioration in the fertilising ability of spermatozoa stored at $-196\,°C$.

Packaging of sperm suspensions to be frozen is in the form of:
1. polyvinylchloride (pvc) straws containing 0.25–0.5 ml;
2. glass ampoules containing approximately 1.0 ml; or
3. pellets containing approximately 0.1 ml.

Disposable straws are now used extensively for bull semen, not least for their convenience at the time of insemination. Pelleting has been used for storage of boar spermatozoa, but thawing and resuspension of spermatozoa in pellets is clearly less convenient than the use of straws or ampoules. In any event, deep-frozen boar semen is not employed widely at the

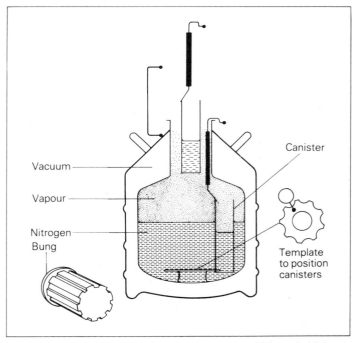

Figure 2.12 Storage cylinder for semen samples which are held deep-frozen at $-196°C$ in liquid nitrogen. Semen may be packaged in glass ampoules or more commonly in polyvinylchloride (pvc) straws which are then used directly in the insemination procedure

present time, partly because the semen of individual boars of proven high fertility differs quite widely in sensitivity to preservation treatments.

Techniques of insemination

Artificial introduction of semen into the female tract aims to deposit the sperm suspension as effectively as the male of the species, and indeed in cattle and sheep the site of insemination is usually deeper in the tract than that found in natural mating. This, of course, is one of the reasons why extensive dilution of the ejaculate is possible without an adverse influence on conception rate. The actual procedure of insemination involves a catheter, pipette or straw, but only in the case of pigs does the catheter bear a close resemblance to the penis. Here the catheter is rubber with a spirally arranged (corkscrew) tip that mimics the boar's glans penis: it is introduced through the vagina into the muscular ridges of the cervix while being rotated in an anticlockwise direction until 'locked' (Fig. 2.13). The site of semen deposition is thus similar to that achieved during mating. Loss of semen from the gilt or sow during actual insemination is a result of contractions of the uterus and an insufficient lock of the catheter in the cervical ridges, but some backflow must be expected after insemination since the gelatinous fraction which normally blocks the cervix has been removed.

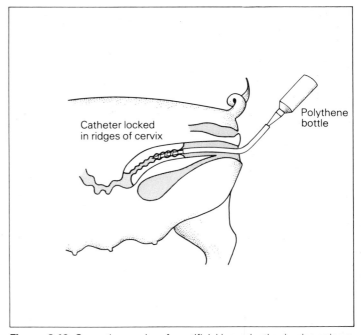

Catheter locked in ridges of cervix

Polythene bottle

Figure 2.13 General procedure for artificial insemination in pigs using a flexible rubber catheter and polythene bottle. The spirally-tipped catheter is introduced along the vagina and into the cervical ridges by an anticlockwise movement, after which the inseminate is delivered gently into the uterus by squeezing the bottle

Insemination in cattle has, for many years, used a glass or plastic pipette that is guided by manual assistance *per rectum* into the uplifted cervical os (Fig. 2.14) and into the mucus-filled canal. In the case of parous cows, the cervix would previously have undergone sufficient growth and stretching to enable the pipette to enter the uterus: in such situations, the procedure was frequently to deposit a proportion of the inseminate in the uterine body and then, by gradual withdrawal of the pipette, to leave the remainder in the cervical canal. However, now that deep-frozen bull semen is stored in plastic straws, these are also used for insemination and are fine enough normally to negotiate the cervix in heifers. The 0.25–0.5 ml semen dose is expelled from the straw by means of an insemination 'gun' (the Cassou gun).

Artificial insemination is not yet widespread in sheep, in part because of the considerable mating potential of individual rams and also because the success rate obtained with frozen-thawed ram semen has frequently been poor. Even so, stored ram semen is available commercially. The technique of insemination involves viewing the cervix by means of a duck-billed speculum that opens the vagina (Fig. 2.15) and introducing the pipette into the cervix; the inseminate can rarely be deposited very deeply into the tortuous canal. The procedure can be completed most conveniently with the ewe on a raised platform, or head downwards in an insemination cradle.

Insemination of horses is little practised in the UK, and is

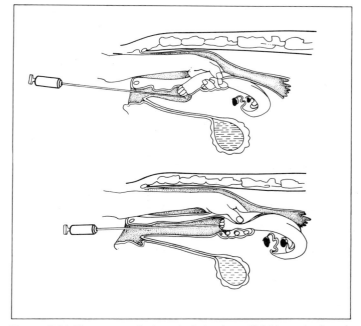

Figure 2.14 The recto-vaginal manipulation for artificial insemination of cattle. The muscular cervix is located through the wall of the rectum, and the cervix is then raised to permit entry of the insemination straw or pipette into the canal – or indeed into the body of the uterus in parous cows

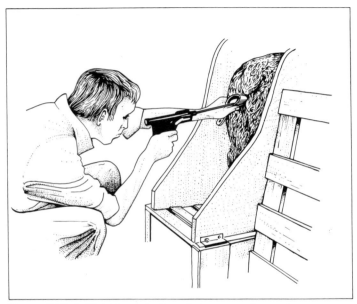

Figure 2.15 Artificial insemination of sheep usually employs a speculum for opening the vagina so that the cervix can be viewed. The inseminate is introduced into the first folds of the tortuous cervical canal

not differ greatly from those already outlined, and the inseminate (usually fresh rather than stored semen) is introduced with a pipette through the relaxed cervix.

The actual success of insemination procedures, as determined by the proportion of females conceiving, depends chiefly on the following four factors:

1. The quality (inherent fertility) of the sample of spermatozoa available.
2. Careful handling of the semen sample, especially with regard to control of temperature, prior to insemination.
3. Insemination at the correct time before ovulation (see Ch. 3, p. 54).
4. Accurate deposition of the inseminate of the female genital tract.

Potential advantages of the technique of artificial insemination are listed in Table 2.3.

currently not accepted by the breeding fraternity for subsequent registration of foals, but the overall procedures do

Table 2.3   Some benefits associated with the use of artificial insemination; these come under the headings of livestock improvement, disease control and economic aspects

1. Enables the widespread use of outstanding sires, and dissemination of valuable genetic material even to small farms
2. Facilitates progeny testing under a range of environmental and managerial conditions, thereby further improving the rate and efficiency of genetic selection
3. Leads to improved performance and potential of the national herd, and permits coordination of breeding policy on a national basis
4. Permits crossbreeding to change the production emphasis, such as from milk to beef
5. Accelerates introduction of new genetic material by export of semen, and reduces international transport costs
6. Enables the use of deep-frozen semen after the donor is dead, thus aiding preservation of selected lines
7. Permits use of semen from incapacitated or oligospermic males
8. Reduces the risk of spreading sexually transmitted diseases
9. Is usually essential after synchronisation of oestrus in large groups of animals
10. Provides a necessary research tool for investigating many aspects of male and female reproductive physiology

Mating, fertilisation and maintenance of pregnancy

3

The act of mating involves behavioural interactions between mature males and females leading, after appropriate stimulation, to mounting, thrusting, intromission and ejaculation. The male usually dismounts after this sequence although – as we shall see – boars may remain engaged in the female tract for several waves of ejaculation. While the primary purpose of mating is to deposit semen in the reproductive tract of oestrous females, it may also influence the timing of ovulation. The evidence for such an influence in farm animals remains equivocal, but a number of fur-bearing mammals such as rabbits, cats and ferrets are *induced ovulators*; the surge of gonadotrophic hormones that programmes collapse of the follicles is triggered by afferent nervous stimuli arising from copulation. As a consequence, a more precise relationship is obtained between introduction of spermatozoa and release of the eggs: failure or anomalies of fertilisation due to ageing of the gametes in the female tract should thus be avoided.

This chapter deals with the events from the time of mating until pregnancy is established and the placenta formed. Sources of prenatal loss will also be considered.

Mating, sites of semen deposition, semen losses

Depending on the system of farming, females are bred by free-mating, hand-mating or artificial insemination. In the first of these, the mature male is left constantly with the females during the breeding season so that – with an appropriate ratio of males to females – he can inspect them continuously, monitor them during pro-oestrus, and then mate them as soon as oestrus is displayed. Repeated matings will usually occur during the period of oestrus although, if several females are in heat simultaneously, the male may show preference for just one of these. Free-mating systems are most common for sheep and beef cattle – in other words, in extensive systems of production.

Hand-mating refers to the method of restricting access of the stud male to oestrous females for quite limited periods. Oestrus is identified either on the basis of behaviour with other females or by checking with a vasectomised teaser male, and she is then mated by the stud male after which he is returned to his pen. This approach is common with bulls and boars, and avoids wasting mating potential or exhausting the male and his semen reserves. Nonetheless, a second or third hand-mating may be considered desirable during the period of heat.

Successful use of artificial insemination relies on correct identification of oestrus, and confirmation of the stage of the cycle is usually possible in cattle on the basis of uterine tone (i.e. the taut and turgid feeling of the swollen uterus during oestrus) and the presence of a large follicle in one of the ovaries; the uterus and ovaries may be palpated *per rectum* prior to introducing the insemination straw or pipette.

During inspection of an oestrous female, males pay particular attention to odours emanating from the vulva and vagina, which seem to be associated with the abundant secretion of mucus at this phase of the oestrous cycle. If the female urinates during inspection, the male also shows interest in this fluid. Olfactory sampling is usually followed in rams, bulls and stallions by a raising of the head, curling back of the upper lip, and a protracted form of sniffing – the Flehmen reaction. Extrusion of the penis from the preputial sheath may occur at this stage, although mounting of the female is sometimes so rapid that penile movements may not be distinguished. Vigorous pelvic thrusting follows until the penis enters between the vulval lips into the vagina and, in rams and bulls, ejaculation is then almost instantaneous with deposition of semen in the anterior vagina and on the external cervical os. The male then dismounts before each repeat mating.

In some contrast, thrusting movements by the boar are more extensive while the spirally arranged glans (tip) of the penis passes along the vagina and actually enters the muscular folds of the cervix. Only when the glans penis is firmly locked in the cervical folds (Fig. 3.1) by means of a twisting motion does thrusting cease and the prolonged process of ejaculation commence. This involves secretion of both a gelatinous and watery pre-sperm phase, followed by a creamy sperm-rich phase, and then gelatinous secretions in a further watery phase. Such a complete wave of ejaculation may take 3–5

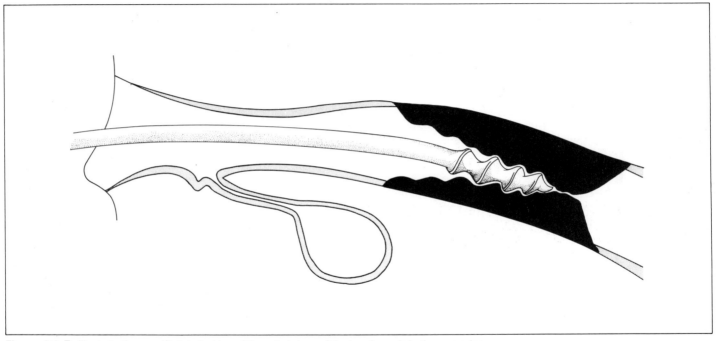

Figure 3.1 To illustrate the specific interlocking of the spiral glans of the boar's penis in the muscular folds of the cervix that provides the stimulus to ejaculation. Because of this arrangement, the ejaculate passes almost directly into the uterus

minutes, and the whole sequence may then be repeated a second or even a third time. Renewed thrusting and establishment of the cervical lock occurs between each sequence. The voluminous ejaculate produced by the boar (see Ch. 2, p. 30) is propelled almost directly into the uterus, although passing contact is made with the innermost portions of the cervical canal (Fig. 3.1).

There is a marked contrast, therefore, between the essentially *intra-uterine* site of semen deposition in gilts and sows and the *intra-vaginal* site in ewes and cows. Artificial insemination advances the site of semen deposition in these ruminant species, into the first cervical folds in sheep and usually into the uterus in cattle. Loss of semen occurs after mating in all three species: in ewes and cows by simple drainage from the vagina, and in pigs by active displacement of semen backwards through the cervix. This may be found during ejaculation when the volume being emitted by the male

is greater than can pass forward at that moment, thus causing leakage. A more marked loss also occurs after dismounting by the boar when contractions of the uterine horns may expel much of the ejaculate. These losses can be reduced by avoiding stress, fright or undue movements at the time of mating or shortly thereafter; a post-coital period of tranquility should therefore facilitate the transport of spermatozoa and establishment of sperm reservoirs in the female tract, ultimately benefiting the conception rate (Table 3.1).

Sperm transport, reservoirs and lifespans in vivo versus in vitro

The prime tactical objective of mating is to enable a population of viable spermatozoa to pass from the ejaculate to the site of fertilisation in the oviducts shortly before the time of ovulation. This allows penetration and activation of the eggs before any deleterious influence of post-ovulatory ageing is found (see below). Because of the different sites of semen deposition in farm animals, certain distinctions in the processes underlying sperm transport are to be expected between ruminants and pigs; even so, several aspects are common to both groups.

First, some spermatozoa reach the oviducts quite rapidly after ejaculation and, considering the distances involved relative to the size and motility of a sperm cell, at a much

Table 3.1 To emphasise the deleterious influence of stress – as seen in behavioural symptoms – on conception rates following artificial insemination of pigs

Behavioural state of female	Total inseminations (no.)	Overall conception (%)
Very calm	875	63.8
Calm	2 625	52.3
Agitated	805	34.1

greater rate than could be accounted for by swimming activity alone. The inference here is that contractions in the smooth muscle of the female genital tract assist transport of spermatozoa and, with the exception of the situation during parturition, the uterus is at its most active following coital stimulation. In fact, contractions may be further magnified by smooth muscle stimulating agents in semen – for example, prostaglandins, which are rich in the ram ejaculate. Moreover, in the absence of fright or other forms of stress, mating causes a reflex release of oxytocin from the posterior pituitary gland, and this hormone acts to enhance the contractions of the female system. Although its effect is only pronounced for a few minutes after release, it is thought to play an important role in sperm transport, especially in assisting a vanguard of spermatozoa up to and into the oviducts. As oxytocin is antagonised by a release of adrenalin, conditions of stress must be avoided. Despite this emphasis on contractile activity, sperm motility is critical in certain regions of the tract such as the cervix and utero-tubal junction and during penetration of the egg membranes.

Second, only a minute fraction of the population of spermatozoa deposited at ejaculation reaches the oviducts at any one time, and indeed only a very small proportion of spermatozoa in the ejaculate ever reaches the oviducts: the majority of sperm cells are lost by voiding or engulfment by white blood cells (polymorphonuclear leucocytes). The inference here is that a gradient in the density of spermatozoa is found along the length of the female tract, with the highest density at the site of deposition and the lowest density in the upper regions of the oviducts (Fig. 3.2). This accurately reflects the situation for the first hours after mating, but since there is a marked loss of spermatozoa from the site of deposition, the highest density is soon found in the reservoir region. Spermatozoa have actively migrated into reservoirs from the neighbouring pool of ejaculate, these regions being the cervix in the cow and sheep and the utero-tubal junction and isthmus in the pig. These structures contain the major reserves of spermatozoa after those at the site of ejaculation have been lost. The reservoirs function as barriers to sperm transport in the sense that they restrict the numbers of spermatozoa passing to the higher regions of the tract, and in this respect they can be viewed as having a regulatory function.

As already noted, the procedure of artificial insemination may overcome part of the barrier function of the cervix in the cow by introducing some or all of the inseminate into the uterus. This enables a massive dilution of the ejaculate (see Table 2.2, p. 36) and, indeed, were this not performed prior to intra-uterine insemination of cows and ewes, there would be a considerable risk of abnormal fertilisation. In addition to providing a regulated supply of spermatozoa to the oviducts throughout the period of oestrus, the sperm gradient acts to reduce the chances of more than one sperm simultaneously reaching and penetrating the egg membranes. Simultaneous

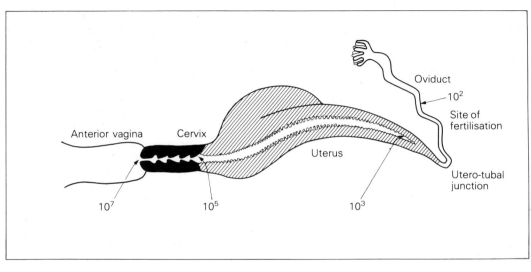

Figure 3.2 To show the diminishing gradient in sperm numbers between the site of ejaculation in the anterior vagina of cows and sheep and the site of fertilisation in the oviducts. This gradient in sperm numbers reduces the chance of too many spermatozoa reaching the egg surface simultaneously, thereby helping to avoid abnormal fertilisation

penetration by two or more spermatozoa – the condition of polyspermy – is lethal in mammals and prevents formation of a normal embryo. As will be discussed later, polyspermy is a frequent sequel to delayed insemination when a sperm gradient may not be adequately imposed.

A third generalisation concerning sperm transport is that the ejaculated cells are progressively removed from the seminal plasma (the fluid component of semen) as they ascend the female tract, and they become resuspended in the secretions of the uterus and then in those of the oviducts. The principal removal of seminal plasma occurs during passage through the mucus-filled cervical barrier in ruminants and the utero-tubal junction of the pig. Resuspension of the ascending spermatozoa is associated with their final maturation (i.e. capacitation) and an increased respiration and swimming activity. Spermatozoa introduced artificially into the oviducts in the presence of seminal plasma cannot fertilise eggs for a period of hours since the male secretions in some way stabilise the spermatozoa and prevent their final ripening. A coordinated loss of seminal plasma is thus necessary during the process of sperm transport, such that the activity and fertilising potential of sperm cells come under the control of the luminal fluids in the female tract. As a corollary, spermatozoa still in contact with seminal plasma constituents in the reservoir regions may have both their metabolism and cell membranes stabilised – conditions that would seem necessary for prolonged viability during storage.

There is still some controversy as to the speed with which ejaculated spermatozoa reach the oviducts in sheep and cattle, and also whether dead cells can be transported to this region. Judgements on the latter point are frequently confusing since it is difficult to know whether dead spermatozoa recovered from the oviducts were dead on arrival or had succumbed subsequently. Recent evidence derived from surgical experiments on sheep indicates that 8–10 hours are required after mating for sufficient spermatozoa to enter the oviducts to ensure fertilisation, and comparable figures are found for cattle. These studies also suggest that spermatozoa are transported to the site of fertilisation more rapidly at the end of oestrus than at the onset – probably reflecting a slow initial build-up of sperm numbers followed by increased contractile activity of the female tract close to the time of ovulation. In contrast to the timing in ruminants, large populations of spermatozoa are found in the oviducts of pigs within 15–30 minutes of mating, this relatively brief interval reflecting the intra-uterine site of semen accumulation and the fact that the utero-tubal junction is the only barrier to cross before spermatozoa enter the oviducts (Fig. 3.2, p. 47).

Two more specialised aspects of the process of transport concern whether there is selection of a genetically 'fitter' population of fertilising spermatozoa by the female tract and whether spermatozoa are attracted to the site of fertilisation or to the eggs themselves by specific substances – which would indicate the phenomenon of chemotaxis. There is no evidence that spermatozoa are actively 'inspected' and selected by the female tract according to their genotype such that a genetically superior population reaches the eggs. On the other hand, there is evidence that spermatozoa of poor motility and morphology are selected *against* in the sense that they are less likely to pass the barrier regions and reach the site of fertilisation. As to chemotaxis, experiments so far have failed to reveal the existence of this phenomenon in mammals.

Lifespans and loss

Any consideration of sperm lifespans – that is, the period after liberation from the male during which these cells retain viability – is complicated by several factors. First, spermatozoa exhibit swimming ability (motility) much longer than fertilising ability: motility is easily judged whereas fertilising ability and subsequent conception rate are the criteria that interest farmers and animal breeders. Second, the ejaculate is composed of a heterogeneous population of millions of cells – heterogeneous in the sense that some are under-ripe, some are at an optimum stage of membrane maturity, and some are already dying or dead. Accordingly, values for sperm lifespans describe the situation for *populations of cells* rather than portraying it accurately for the relatively small numbers of cells that would be found in a fully diluted ejaculate. Third, sperm lifespans in the female reproductive tract differ from those *in vitro*, especially where deep-frozen storage of cells is being considered. Cooling reduces metabolic rate and deep-freezing may permit an indefinite lifespan (see Ch. 2, p. 36).

Despite these strictures, the fertilising lifespan of spermatozoa in the female falls within a period of 24–48 hours from ejaculation (Table 3.2), with some reduction in fertility as ejaculate age reaches the upper end of this lifespan (Fig. 3.3) or if spermatozoa have been chilled or otherwise damaged during artificial insemination procedures. Since ovulation time is usually within 40 hours of the onset of oestrus in cattle, sheep and pigs, ageing of spermatozoa is unlikely to cause a significant depression of fertility. This statement assumes that a full ejaculate of normal concentration has been deposited and that the female has not been subjected to hormonal treatments that could influence the tract. The risk of compromising fertility due to sperm

Table 3.2 The fertilising lifespan of eggs and spermatozoa in the female reproductive tract

Species	Eggs (hours)	Spermatozoa (hours)
Cow	10–12	30–48*
Sheep	10–15	30–48*
Pig	8–10	24–42*

NB. Sperm motility retained for much longer than fertilising ability.
*In contrast to the estimate for eggs, these figures refer to populations of millions of spermatozoa rather than to individual cells.

ageing may be greater under conditions of artificial insemination where oestrus may not have been detected with certainty (i.e. insemination too soon) and a much reduced population of viable cells is introduced.

When considering the vast numbers of spermatozoa deposited in the female at mating, the question arises as to the ultimate fate of these cells. The majority are lost from the female tract by voiding within an hour or two of mating. Conditions of stress exacerbate this form of loss and thus reduce the numbers of spermatozoa potentially available for transport. The next most important means of loss is through phagocytosis by polymorphonuclear leucocytes; these white cells are liberated into the tract shortly after mating and engulf dead, dying and poorly motile cells as well as bacteria entering with the ejaculate. Polymorphs are also active in the peritoneal cavity, and spermatozoa passing upwards through

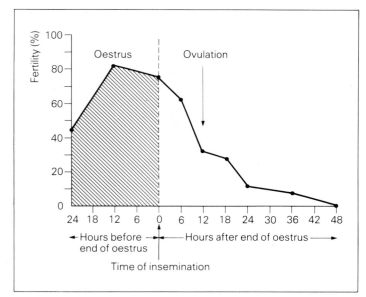

Figure 3.3 The influence of the time of insemination on conception rate in cattle. The optimum time of insemination for maximum fertility is some 12–18 hours *before* ovulation. Late insemination causes a drastic fall in potential fertility

the oviduct and out of its fimbriated extremity likewise suffer engulfment. A third route of sperm disposal, although probably applicable to only a very small proportion of cells in the ejaculate, includes incorporation in the egg membrane (the zona pellucida) and cellular investments (the granulosa cells) and possibly in the epithelium lining the uterus and oviducts. The latter is controversial, for epithelial incorporation of free spermatozoa – as distinct from phagocytosed cells – raises the question of formation of anti-sperm antibodies and reduced fertility.

Ovulation, egg transport and egg ageing

The process of ovulation involves collapse of the fully mature Graafian follicle and release of the egg into the oviduct. Each follicle sheds a single egg, therefore treatments that increase the number of mature follicles – as may occur with nutritional flushing in certain situations – also increase the potential number of fertilised eggs, and thus offspring.

The pre-ovulatory surge of gonadotrophic hormones causes critical changes in the egg itself and in its surrounding cell layers. Most important of these, the chromosomes of the egg are halved by a meiotic division, the egg thereby progressing from a primary to a secondary oocyte ready for sperm penetration (Fig. 3.4). Remarkably, the egg has been resting as a primary oocyte since its formation in the ovaries in early foetal life (Ch. 1, p. 9), and it only resumes meiosis just before release at ovulation. The arrested stage may last for

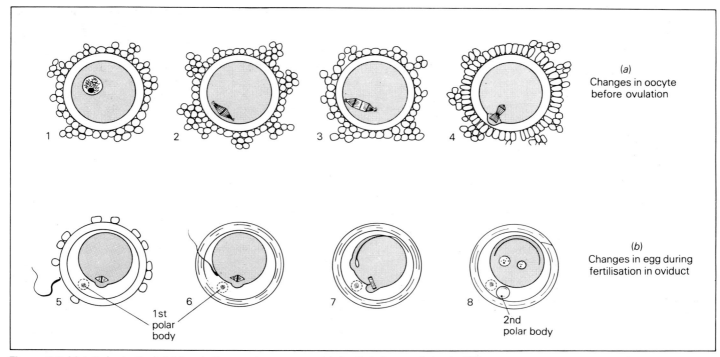

(a) Changes in oocyte before ovulation

(b) Changes in egg during fertilisation in oviduct

1st polar body

2nd polar body

Figure 3.4 Meiotic (reduction) divisions in the egg chromosomes (a) shortly before ovulation and (b) at the commencement of fertilisation leading, respectively, to extrusion of first and second polar bodies

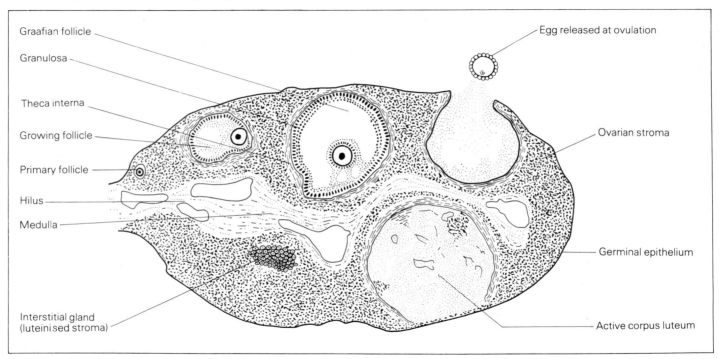

Figure 3.5 Simplified section through an ovary to show growth of a Graafian follicle and then release of the egg as the follicle collapses at ovulation. The collapsed follicle subsequently forms a progesterone – secreting body termed a corpus luteum

several years in farm animals and for 40 or more years in women. The cells that surround the egg become more loosely arranged – a change that facilitates release of the egg from the follicle into the oviduct and probably also aids penetration of the fertilising spermatozoon to the egg surface. Further nuclear changes follow rapidly if the egg is fertilised (Fig. 3.4). Although the follicular fluid that escapes from the antrum at ovulation may carry the egg with it, contractions in the follicle wall after initial collapse help to displace eggs in which cellular connections have not been fully severed (Fig. 3.5). A local release of prostaglandins would promote smooth muscle contractions.

The function of the collapsed follicle is now to form a corpus luteum (plural: corpora lutea) capable of secreting sufficient progesterone to maintain a pregnancy. This is achieved by multiplication of the granulosa and sometimes theca interna cells (hyperplasia) and also by an increase in their size (hypertrophy) such that a solid mass of steroid-secreting luteal cells is built up within five to six days of ovulation.

Egg transport

In contrast to the situation in poultry where eggs may occasionally enter the body cavity and cause peritonitis, mammalian eggs are invariably conveyed safely from the surface of the follicle into the ostium of the oviduct and rapidly along the thin-walled ampulla to the site of fertilisation at the ampullary-isthmic junction (Fig. 3.6). Two factors in particular contribute to this phase of egg transport from the gonad into the reproductive tract: first, the unfertilised egg is still invested in a dense layer of follicular cells which, by rearrangement just before ovulation, surround the egg with finger-like processes. Second, the funnel-shaped fimbriated end of the oviduct completely covers the ovary at ovulation, and its inner surface is lined with cilia beating downwards into the ampulla. Together with the action of peristalsic contractions, the cilia ensure that the egg(s) is transported to the ampullary-isthmic junction within 30–45 minutes of ovulation. In mated animals, the egg should thus encounter spermatozoa very soon after its release from the follicle, although the initial period of exposure (about half an hour) to the oviduct environment may be necessary for a final maturation of the egg membranes. Passage of the egg, its follicular cells and/or some follicular fluid into the oviduct stimulates sperm transport from the isthmus to the site of fertilisation, thereby ensuring a prompt penetration of the eggs.

Egg ageing

The point has already been made that many different stages of ripeness may be present within the ejaculate since the sperm population is a heterogeneous collection of cells; paradoxically, this very heterogeneity increases its fertilising

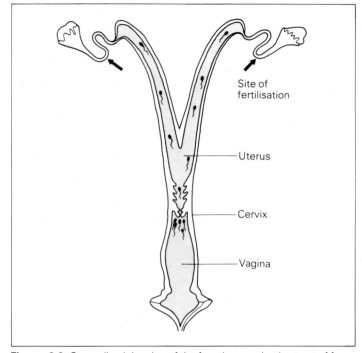

Site of
fertilisation

Uterus

Cervix

Vagina

Figure 3.6 Generalized drawing of the female reproductive tract of farm animals (cow, sheep) to show the site of fertilisation in the oviducts and migration of spermatozoa to this region

lifespan. By contrast, nearly all (if not all) eggs ovulated spontaneously are initially fully fertilisable, the interpretation being that follicles mature enough to ovulate spontaneously will release viable eggs. However, post-ovulatory ageing of unfertilised eggs is a fairly rapid phenomenon, degenerative changes first being detected within 8–10 hours; such changes are quite distinct within 12 hours of ovulation (Table 3.3). The

Table 3.3 To show the effects of post-ovulatory ageing of eggs in the oviducts on fertilisation and on embryonic survival 25 days after insemination in pigs. Eighteen animals were inseminated in each group

Estimated age of eggs at fertilisation (hours)	Eggs fertilised normally (%)	Viable embryos at 25 days (mean)
0 (control)	90.8	12.0
4	92.1	11.7
8	94.6	8.7
12	70.3	6.8
16	48.3	4.8
20	50.9	5.0

NB. Note the loss in the 8-hour group of fertilised-aged eggs by day 25 of gestation.

message is therefore clear: mating or artificial insemination must occur *before* ovulation if the risk of egg ageing is to be avoided. In fact, studies in cows, sheep and pigs indicate that the optimum time for mating or insemination is 12–14 hours

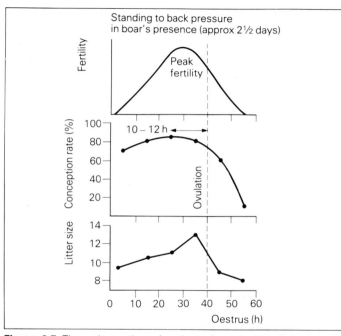

Figure 3.7 The optimum time of mating or insemination of pigs in order to obtain maximum conception rate and litter size. If the time of ovulation is known or can be judged accurately, then the optimum time for semen deposition is 10-12 hours before ovulation or early on the second day of oestrus

before ovulation (Figs 3.3 (p. 50) and 3.7). Use of this information assumes that the time of ovulation is known or controlled (see Ch. 6, p. 110).

The influence of post-ovulatory ageing of eggs is expressed in lower conception rates in cattle and sheep (Fig. 3.3) and, in addition, a smaller litter size in pigs (Table 3.3). As eggs age in the oviduct while awaiting the arrival of spermatozoa, they undergo changes in their membranes, cytoplasm and nucleus which together lower their fertilisability and/or potential to yield a viable embryo. Post-ovulatory ageing of eggs may therefore cause lowered fertility not only due to eggs failing to be fertilised but also to embryonic mortality following fertilisation of ageing eggs. One of the more conspicuous deleterious changes is in the *block to polyspermy*: ageing eggs are less able and act more slowly to prevent penetration of the cytoplasm by two or more spermatozoa, with the resultant polyspermy causing early death of the embryo. Cytoplasmic organelles also wander from their functional locations during a period of ageing, whereas nuclear changes include loss of one or more chromosomes from the meiotic spindle – once more a condition that is invariably lethal should a zygote subsequently be formed.

Fertilisation and embryonic development in oviduct

As noted above, ejaculated spermatozoa have to undergo a final phase of maturation in the female reproductive tract before they can penetrate the egg membranes. This process is referred to as *capacitation* – the acquisition by spermatozoa of the capacity to fertilise; it requires 1½–5 hours, according to species (Table 3.4). Spermatozoa increase their swimming

Table 3.4 Timing of capacitation. The figures represent the period required by spermatozoa in the female tract before they are competent to penetrate the egg membranes

Species	Interval (hours)
Bull	4–5
Ram	1–1.5
Boar	2–3

and respiratory activities upon capacitation, and the incisive progressive motility is considered essential for fertilisation. As a consequence, however, capacitated spermatozoa are short-lived cells whose potential is replaced by others proceeding along the curve of ripening and senescence.

Membranous changes follow immediately upon capacitation, especially on the anterior portion of the sperm head; their function is to enable release of the enzymes used in penetration of the egg investments. Two enzymes contained in the acrosome (Fig. 3.8) are thought to be essential for fertilisation: (1) hyaluronidase to depolymerise a pathway for the spermatozoon in the hyaluronic acid cement between the follicular cells that surround the egg; and (2) acrosin to enable digestion of a slit through the thick mucopolysaccharide membrane, the zona pellucida, that envelopes each egg. The enzymes are released by means of a process of fusion and vesiculation of the outermost two membranes of the sperm head – the plasma membrane and the outer acrosomal membrane (Fig. 3.8). The *acrosome reaction* thus enables lytic enzymes to escape through ports in the membranes. The very fact that this process follows automatically once capacitation is achieved is another reason why capacitated spermatozoa rapidly become non-functional, due to loss of their lytic enzymes.

Immediately after traversing the zona pellucida and entering the perivitelline space, the sperm head fuses with the egg plasma membrane, activity of the sperm tail ceases, and the head and most of the tail are incorporated into the egg cytoplasm. Fusion of the sperm head with the egg surface causes *activation* of the egg (secondary oocyte). Activation takes three forms:

1. A second reduction division, with extrusion of the second polar body and formation of a haploid set of female chromosomes (Fig. 3.4, p. 51).
2. Fusion of cytoplasmic organelles, termed 'cortical granules', with the egg plasma membrane, enabling release

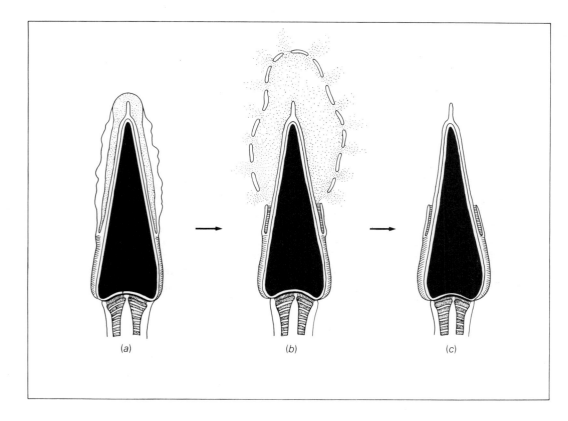

(a)   (b)   (c)

Figure 3.8 The acrosome reaction – the vesiculation reaction around the anterior portion of the sperm head whereby the outermost membranes swell and fuse in order to release enzymes for penetration into the egg

of their contents and instigation of the block to polyspermy (Fig. 3.9). Cortical granule contents act enzymatically to change the nature of the zona pellucida and egg plasma membrane. These changes together prevent further spermatozoa entering the egg cytoplasm, although the changes are slower or non-functional in ageing eggs.

3. Stimulation of synthetic activity in the cytoplasm and nuclei, including replication of the DNA content of the male and female chromosomes.

The two haploid groups of chromosomes each become surrounded by a nuclear membrane, and are then termed 'pronuclei'. Their chromosomes diffuse as threads of chromatin, and the pronuclei migrate to meet each other in the centre of the egg. DNA replication is most active during this migratory phase so that, upon meeting of the two pronuclei and condensation of the chromosomes, these can reassemble at metaphase on the first mitotic spindle and pass directly to anaphase and telophase. Division of the cytoplasm gives a 2-cell egg, and subsequent cleavage divisions follow according to a strict schedule of developmental events. The first division occurs about 15–20 hours after sperm penetration into the secondary oocyte.

As discussed previously, abnormalities of fertilisation are most common in ageing eggs. As well as the condition of polyspermy, there may be failure of extrusion of the second polar body, usually as a result of some rotation of the meiotic spindle away from the egg surface. Retention of the second

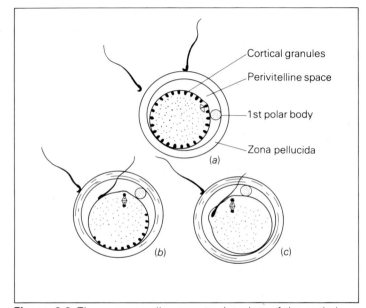

Figure 3.9 Three mammalian eggs to show loss of the cortical granules that occurs as a response to sperm penetration. The cortical granule material diffuses across the fluid-filled perivitelline space to make the zona pellucida impermeable to further sperm penetration – the block to polyspermy

polar body causes the egg to have two haploid sets of female chromosomes (digyny) which, together with the haploid set introduced by the spermatozoon, will constitute a triploid embryo – a lethal condition. Digyny and polyspermy may arise in the same egg. Other anomalies of the chromosome complements and fragmentation of the cytoplasm are also found.

Early development

The embryos remain in the oviduct for two days in the pig and a little over three days in sheep and cows, entering the uterus at the 4-cell stage in pigs and as 8–16 cells in sheep and cattle (Fig. 3.10). Transport through the isthmus of the oviduct is regulated principally by the influence of ovarian steroid hormones on the circular and longitudinal muscle layers in the wall. The increasing concentration of progesterone from the developing corpus luteum promotes a gradual relaxation of the musculature, a reduction in oedema in the mucosa, an increase in the size of the oviduct lumen, and contractions orientated towards the uterus. The flow of tubal fluid may also contribute to the transport of embryos into the uterus.

Although the eggs of farm animals are released from the ovary with considerable cytoplasmic reserves of yolk, the developing embryo very soon becomes dependent for its sustenance on fluids of the reproductive tract – a situation which changes only when the placenta is formed and there is

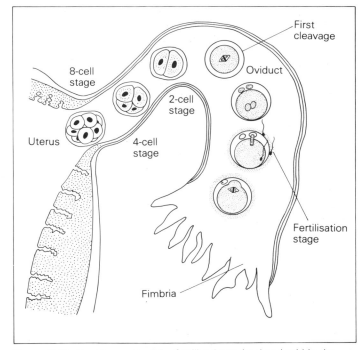

Figure 3.10 Development of the fertilised egg (embryo) within the oviduct so that it has achieved 4-8 cells or more by the time it enters the uterus 2-3 days after ovulation

access to the maternal blood supply. In the oviduct, the volume of the luminal fluid varies with the stage of the oestrous cycle and is greatest at or shortly after ovulation when gametes and embryos would be present (Fig. 3.11). A further point is that the ability of the embryo to use substrates in oviduct fluid varies with its stage of development, and the available concentration of substrates such as Kreb's cycle intermediates is also changing. For example, extensive studies based on mice – from which large numbers of embryos may be obtained inexpensively – have shown that single-cell embryos can incorporate only pyruvate and oxaloacetate, whereas 2-cell embryos can incorporate pyruvate and lactate, and 8-cell embryos malate and, indeed, glucose. There is thus a dynamic relationship between the developing embryo and its fluid environment.

This phase of embryo development and transport in the oviduct is particularly susceptible to disruption by changes in the balance of circulating steroid hormones, as might be found after treatments for synchronising oestrous cycles (see Ch. 6, p. 104) or after ingestion of oestrogenic plants. For just as ovarian steroids control the muscular activity of the oviducts, so they control secretory activity in both a quantitative and qualitative sense.

While description of developmental events quite rightly focuses on the oviducts for the first two or three days after ovulation, the uterus is also changing under the influence of ovarian progesterone; the changes are essential for successful

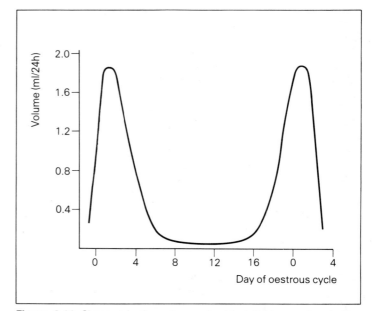

Figure 3.11 Changes in the volume of oviduct fluid according to the stage of the bovine oestrous cycle. The volume of fluid accumulating (both as secretion and as transudate) in all species examined is greatest at the time of and shortly after oestrus when gametes and embryos would be present in the oviducts

reception of the embryos. First, and foremost, after the end of oestrus there is a major invasion of polymorphonuclear leucocytes whose function is to cleanse and sterilise the uterine lumen by ingesting dead and dying spermatozoa, seminal waste and bacteria introduced at the time of mating. The glandular epithelium of the uterus (termed the 'endometrium') also proliferates during this interval and secretory activity likewise increases, enabling the embryo to move from the nutritionally important fluids of the oviduct to those of the uterine lumen; the latter are sometimes termed 'uterine milk' or 'histotrophe'. One interpretation of the 2–3 day sojourn in the oviduct is not simply for the embryos to benefit from that environment but also to give the uterus time to respond to increasing concentrations of progesterone from the developing corpus luteum.

As already suggested, the timing of embryo entry into the uterus is critical for further development. Much experimental work in laboratory and farm species, especially using the technique of embryo transplantation, has shown that a precocious *or* retarded passage of the embryo into the uterine lumen compromises development, largely due to the embryo being presented to an inappropriate fluid environment. Synchrony between the stages of embryonic and endometrial development is therefore vital for successful transplantation of embryos, and this aspect is discussed further in Chapter 6.

Embryonic development in uterus, attachment and establishment of pregnancy

Embryos continue to develop by mitotic division in the uterus, and by the time the embryo is a ball of 16 or more cells, it is termed a *morula* (Fig. 3.12). Individual cells or blastomeres soon begin to secrete fluid into the intercellular spaces, after which they are rearranged around a central fluid-filled space, the blastocoele. The embryo is now a *blastocyst*, and it is at this stage that the group of cells destined to form the embryo proper (the inner cell mass) becomes distinguishable from those that will form the embryonic membranes (the trophoblast). The blastocyst stage is reached 5–6 days after fertilisation in pigs, and a day or two later in sheep and cattle (Table 3.5).

Table 3.5 Stages of early embryonic development presented as times after ovulation or sperm penetration of the eggs. Pre-ovulatory mating or insemination has been assumed

Species	2 cells (hours)	4 cells (hours)	8–16 cells (hours)	Loss of zona pellucida from blastocyst (hatching) (days)	Beginning of attachment (days)
Cow	20–24	32–36	72–84	9–11	22
Sheep	16–18	28–30	66–72	7–8	15
Pig	14–16	20–24	60–72	6	13

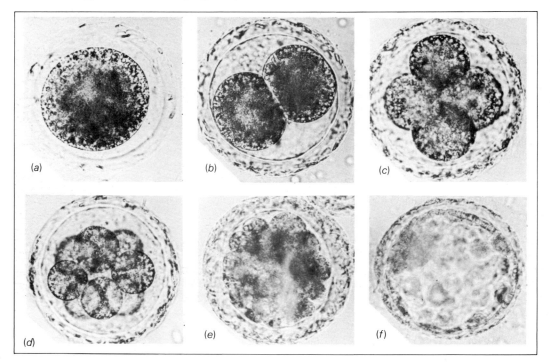

Figure 3.12 The first stages of development of the embryo from the newly-fertilised single-cell egg through the 2-, 4-, 8- and 16-cell stages to that of a blastocyst with a fluid-filled cavity. Note that the zona pellucida still surrounds the embryo and has trapped large numbers of spermatozoa

Further development of the blastocyst becomes limited by the presence of the zona pellucida, and this protective membrane is then shed in a process called 'hatching'. The zona pellucida may be softened enzymatically for emergence of its embryo, but one component of hatching is purely physical: active pumping of water into the blastocoele leads to rupture of the zona pellucida and escape of the embryo (Fig. 3.13). This occurs late on day 6 in pig embryos, on days 7–8 in sheep and days 9–11 in cows.

Before implantation commences, a further developmental process conspicuous in farm animals is a massive elongation of the embryo by proliferation of the trophoblast and rearrangement of the cell layers (Fig. 3.14). In fact, this dramatic transformation is common to the ungulates as a group. Its specific function is unknown, but an extensive form of surface contact between embryo and endometrium may be essential for maintenance of pregnancy before a functional placenta is formed. Elongation is most pronounced in the pig where the trophoblast may attain a length of some 300 cm or more, even though this is never displayed in linear fashion: rather, the embryo is arranged in a zig-zag manner (Fig. 3.14). It is during and briefly after this stage of elongation in pigs that a mixing and redistribution of embryos throughout the continuity of the two uterine horns takes place. This redistribution (*not* active migration) is due to uterine contractions, possibly in response to trophoblastic secretion of hormones. Such a redistribution of embryos compensates for

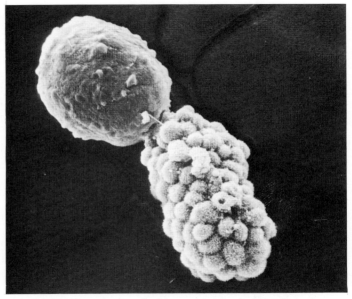

Figure 3.13 Pig embryo at the blastocyst stage hatching from the zona pellucida by expansion, due to active pumping of water into the blastocyst cavity – the blastocoele. (Scanning electron micrograph, courtesy of Dr J. E. Fléchon)

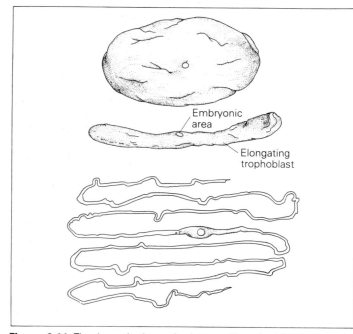

Embryonic area

Elongating trophoblast

Figure 3.14 The dramatic change in shape and length of the pig embryo that occurs between 8-12 days of age, a few days after hatching from the zona pellucida. Embryos are redistributed throughout the two uterine horns during this phase of elongation shortly before commencement of attachment

differences in the number of eggs released from the two ovaries – which may be as dissimilar, for example, as 18 and 3 – and ensures a fairly even spacing before attachment of embryos commences. If redistribution did not occur, overcrowding and embryonic death would inevitably follow. Redistribution of embryos is infrequent in ruminants (<10%).

Further development of the embryo, which in its elongated form is frequently termed a *conceptus*, leads to differentiation of four distinct membranes: (1) the amnion; (2) the allantois; (3) a yolk sac soon to become vestigial; and (4) the surrounding chorion – a derivative of the trophoblast (Fig. 3.15). The allantois expands to associate closely with the chorion, thus forming an allanto-chorion, and the embryo itself is situated within the amnion. These membranes can be distinguished as fluid-filled sacs in farm animals by early in the fourth week of pregnancy. Together, the layers constitute the embryonic or foetal placenta, and function to enable metabolic exchange with the maternal components of the placenta. It therefore follows that the embryo and its membranes already have a developing vascular system, and in fact a heart beat can be distinguished within the third week of life.

Attachment

Attachment of the embryonic membranes to the epithelium of

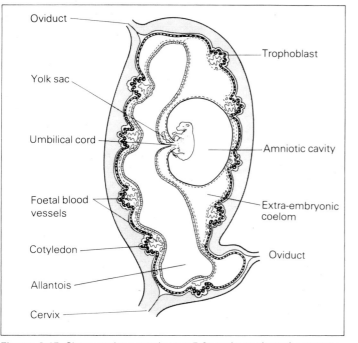

Oviduct

Yolk sac

Umbilical cord

Foetal blood vessels

Cotyledon

Allantois

Cervix

Trophoblast

Amniotic cavity

Extra-embryonic coelom

Oviduct

Figure 3.15 Sheep embryo aged some 5-6 weeks to show the arrangement of the placental membranes, especially the amnion and allantois, and their relationship to the trophoblast or chorion that makes contact with the wall of the uterus

the uterus, the endometrium, is a progressive process but recent studies with the electron microscope have revealed a much earlier start to this intimacy than was previously suspected. Attachment of the pig embryo begins within 13–14 days of fertilisation, that of the sheep embryo 1½–2 days later, and in the cow by 22 days after fertilisation. Because the conceptus remains in the lumen of the uterus in these farm species and does not invade the stroma, attachment is a more appropriate term than implantation. In pigs, the contact between embryo and uterus remains simple, leading to an *epithelio-chorial* placenta, whereas in sheep and cattle there are specialised areas of attachment. These are the maternal caruncles into which the cotyledons of the allanto-chorion interdigitate to form a *syndesmo-chorial placenta*. The attachment complex is termed a *placentome* (Fig. 3.16), and is the area of functional exchange between the two sides of the placenta in ruminants. While the caruncles have an overall convex shape in the cow, they are dish-shaped or concave in sheep, but in both species they contain numerous crypts into which the allanto-chorion interdigitates. Despite the intimacy of the contact, there is no exchange of blood across the placenta; rather, there is diffusion and/or active transport of specific components, with the nature of placental exchange being modified according to the stage of gestation.

Confusion sometimes arises from use of the terms 'embryo' and 'foetus'. Strictly speaking, an egg is an embryo from the completion of fertilisation, so that 2-cell eggs, 4-cell eggs,

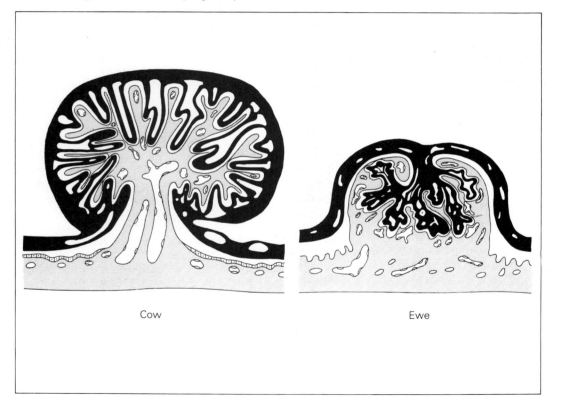

Cow

Ewe

Figure 3.16 The placentomes of cows and sheep – the specialised region of the placenta in which the foetal cotyledons and maternal caruncles interdigitate for the exchange of nutrients and waste products

morulae and blastocysts are all embryos. The more difficult question is 'When does an embryo become a foetus?' Reproductive physiologists regard the embryonic stage as continuing until differentiation of the organ systems is complete – that is, when head, eyes, heart, liver, limb-buds, etc. can be clearly distinguished (Fig. 3.17). Once such differentiation has occurred, the tactical problem then shifts to one of growth of the organ systems, and this is the stage at which an embryo can be considered as a foetus. The foetal stage is first found in farm animals between the fourth and fifth weeks of gestation.

Establishment of pregnancy

A fundamental question to be answered at this point is how does an animal realise that she has conceived and has one or more viable embryos growing in the uterus? In other words, what event(s) causes the animal to cease exhibiting oestrous cycles and, instead, to maintain the corpus luteum and thereby secretion of progesterone to maintain the pregnancy? Reasoning from first principles, there would seem to be at least two possible answers. In a positive sense, the embryo might signal its presence to the mother and cause the corpus luteum (or the many corpora lutea, as in pigs) to be maintained and indeed to increase secretion of progesterone. Alternatively, in the absence of a successful mating, the non-gravid uterus might act to cause regression of the corpus

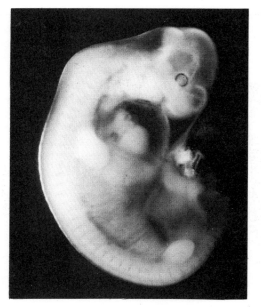

Figure 3.17 A very young pig foetus, aged approximately 25 days, dissected free of its membranes to show substantial development of the major organ systems. Head, eyes, heart, limb buds and spinal column can all be clearly distinguished

luteum, permitting maturation of a follicle(s) and resumption of a new oestrous cycle. Or perhaps the corpus luteum may have a built-in lifespan and, in the absence of one or more viable embryos in the uterus, it may commence to regress automatically.

Much fundamental research during the last ten years has clarified the mechanisms underlying the maternal recognition of pregnancy. Before discussing these points, it is essential to note that the decision whether conception has or has not occurred is taken relatively late after mating and, in terms of the duration of the oestrous cycle, very close to the end of the luteal lifespan (Table 3.6). This can be interpreted as giving

Table 3.6 The duration of the luteal phase of the oestrous cycle in relation to cycle length, recognition of pregnancy, and the duration of pregnancy

Species	Oestrous cycle	Luteal phase	Recognition of pregnancy	Gestation length*	
				Mean	Range
Cow	21	17–18	16–17	282	275–290
Sheep	16–17	14–15	12–13	148	144–153
Pig	21	15–16	12–13	114	111–119

*Usually shorter with multiple births.

the embryo(s) the maximum chance to express its presence before the decision is taken. However, in terms of the duration of gestation, this time is very early and remarkably close to the commencement of embryonic attachment, especially in the pig. There is no specific evidence so far that the attachment

process itself is part of the primary signal for pregnancy maintenance.

Although neurally transmitted information may be important in the establishment of pregnancy, the current understanding focuses on hormonal mechanisms, especially those involving the uterus, ovaries and the anterior pituitary gland. As discussed in Chapter 1, the luteal phase is terminated due to uterine secretion of a hormone, prostaglandin $F_{2\alpha}$ ($PGF_{2\alpha}$). This reaches the ovary largely through a local vascular transfer mechanism and, because it causes regression of the corpus luteum and cessation of progesterone secretion, it is referred to as a *luteolytic hormone*. One demonstration of the luteolytic role of uterine $PGF_{2\alpha}$ comes from hysterectomy – surgical removal of the uterus. If this is performed during the luteal phase, the lifespan of the corpus luteum is postponed indefinitely. A more specific demonstration of the role of $PGF_{2\alpha}$ involves treating cycling animals with antibodies against this luteolytic hormone: the result is a cessation of oestrous cycles and maintenance of the corpus luteum.

$PGF_{2\alpha}$ is released into the uterine veins of unmated animals (Fig. 3.18) at the end of the luteal phase in peak amounts, probably due to an influence of ovarian oestrogen secretion from developing follicles. By contrast, in animals containing one or more viable embryos in the uterus at a time corresponding to the end of the luteal phase, secretion of $PGF_{2\alpha}$ into the uterine veins is largely prevented. This may be

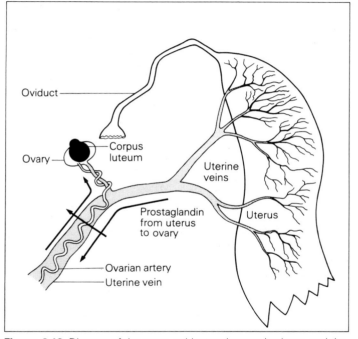

Figure 3.18 Diagram of the ovary, oviduct and uterus in sheep, and the associated uterine veins and ovarian artery through which PGF$_2\alpha$ passes from a non-gravid uterine horn to the ovary to destroy the corpus luteum at the end of the luteal phase

either by actual *suppression* of any elevated release, as seems to occur in ruminants, or by *redirection* of PGF$_2\alpha$ release into a reservoir in the uterine lumen – as is thought to occur in pigs. In either case, increased concentrations of PGF$_2\alpha$ do not reach the ovary and the embryo is therefore said to be exerting an *anti-luteolytic* effect. This is the primary influence of the embryo in maintaining pregnancy – by negating the mechanism that operates during the cycle to terminate progesterone secretion. That this mechanism is not brought into play until the late luteal phase can be shown by transplantation of embryos to the uterus of sheep and cattle shortly before expected luteal regression; with good techniques, a high proportion of pregnancies can be established.

The stage at which pregnancy should be recognised (Table 3.6) also coincides with one of the peaks of embryonic loss in farm animals. Why should this be? In the context of the facts just presented, there is a delicate hormonal balance between the luteolytic action of a non-gravid uterus and the anti-luteolytic activity of the embryo. For example, in cattle and sheep with a single ovulation, slight retardation in development of the embryo may mean that it is unable to exert a sufficiently anti-luteolytic effect at the end of the luteal phase, thereby permitting regression of the corpus luteum and precipitating its own death. In pigs, an insufficient number of viable embryos (seemingly <5), or unequal spacing of embryos leaving a large proportion of one uterine horn unoccupied, will lead to the same fate.

Shortly after this critical phase of anti-luteolytic versus luteolytic activity, the embryo acts in a more positive manner to signal its presence and to stimulate ovarian output of progesterone. This is achieved by secretion of hormones, thought to be synthesised in the trophoblast. Embryonic hormones that act in this manner, particularly to stimulate luteal secretion of progesterone, are termed *luteotrophins*. They may be protein hormones such as chorionic gonadotrophin and trophoblastin, or steroid hormones such as oestrogens. The corpus luteum of pregnancy also receives systemic support in the form of luteotrophic hormones from the anterior pituitary gland: these usually act as a complex of hormones, such as prolactin and LH or FSH and LH. Any factor that interferes with the luteotrophic stimuli, especially with the pituitary luteotrophins, will compromise the maintenance of pregnancy and reduce reproductive performance. For example, in poorly nourished cattle in the depth of winter, there is a diminution or withdrawal of pituitary luteotrophin leading to resorption of the embryo or abortion of the foetus.

While progesterone is essential for the maintenance of pregnancy in mammals, ovarian progesterone may not be the sole supply of a quantity of hormone sufficient for this purpose. Overlooking the limited secretion of progesterone by the adrenal glands, the placenta may develop the ability to synthesise progesterone – although any such output would normally be acting to complement ovarian progesterone. Among the farm species, the sheep placenta is able to secrete sufficient progesterone by day 50 of gestation to maintain the pregnancy even in the absence of the ovaries. Enough has been said to indicate, therefore, that not only is the placenta the organ of metabolic exchange between mother and foetus, but also that it is an extremely versatile endocrine organ. Current evidence indicates that the endocrine activity is derived primarily from the embryonic and foetal – not maternal – placenta.

Embryonic and foetal mortality

While discussion so far has concentrated on the development of embryos and the establishment of pregnancy, not all eggs released at the time of mating or insemination appear as viable foetuses at term. In fact the loss, referred to as *prenatal loss*, is of the order of 30–40 per cent in farm animals and may be increased in certain circumstances. Much of this loss is attributable to genetic causes, is unavoidable, and should be regarded as a normal way of eliminating unfit genotypes in each generation. In polytocous (litter-bearing) species such as pigs, the extent of the prenatal loss is calculated by expressing the number of embryos or foetuses as a percentage of the number of corpora lutea: the latter does not change during

Table 3.7 To illustrate the extent of embryonic and early foetal loss in two groups of sows as judged from the number of embryos or foetuses in comparison with the number of corpora lutea

Stage of gestation (days)	Number of animals	Overall loss of embryos (%)	Average number of surviving embryos or foetuses
25	48	31.8	13.1
40	37	40.2	11.2

gestation (Table 3.7). However, in breeds of cattle and sheep with a single ovulation (i.e. essentially monotocous species), death of the embryo or foetus terminates the pregnancy, and in due course the animal returns to heat. In this situation, therefore, prenatal death must be calculated in populations of animals. Fertility is more commonly expressed in cattle as the *non-return rate*, and may be quoted at various intervals after mating or insemination (Table 3.8). The gradual decrease in

Table 3.8 Proportion of cows considered pregnant at increasing intervals after service or artificial insemination on the basis of non-return to oestrus

Interval from service (months)	Mean (%)	Range (%)
1	72.9	67.2–78.4
2	67.6	58.4–74.8
3	63.8	55.7–70.2
5	60.3	55.1–69.7
Calving	53.5	52.0–57.5

the mean non-return rate is due to prenatal death, and from the table it is again seen that there is approximately a 40 per cent loss by calving time.

What is the distribution of prenatal loss? Under conditions of correctly timed mating or insemination, the incidence of fertilisation in farm animals is close to 100 per cent, but abnormalities may arise in the first week or two of development and differentiation of the embryo. Overall, more than two-thirds of the prenatal loss is found during embryonic development and less than one-third during the foetal stage. Particularly susceptible phases in the former are formation and hatching of the blastocyst around days 6–9 and commencement of attachment from days 13–22. As noted, retarded development of the embryo by the end of the luteal phase may prevent it acting against the uterine luteolytic mechanism; thus, secretion of $PGF_{2\alpha}$ may bring about failure of the pregnancy even though one or more embryos are present in the uterus. If development is in some manner defective, either in terms of timing or on the grounds of abnormal differentiation, it would seem an appropriate biological strategy to eliminate such embryos at an early stage rather than commit the mother to a wasteful and extensive period of gestation before elimination. The reasons for the small proportion of foetal loss are uncertain, but are thought to be associated with uterine constraints (e.g. overcrowding) or with inadequacies of the placenta rather than to factors intrinsic to the foetus.

Discussion of prenatal loss has so far referred to the 'natural' situation, but there are factors that will exacerbate the extent of the loss, such as nutritional stress (e.g. energy shortages, mineral imbalance or vitamin deficiencies), diseases of the reproductive system or endocrine imbalances associated with ovarian cysts. Putting factors of this nature on one side, the single most common cause of embryonic loss is ageing of the gametes in the female reproductive tract due to mating or insemination at the wrong stage relative to heat or, more correctly, to ovulation. For reasons already explained, delayed mating or insemination will prove a more frequent and serious problem than premature insemination, and the deleterious effects of post-ovulatory ageing of the eggs on fertility are most clearly expressed in a polytocous species (Table 3.3, p. 54). The necessity to avoid ageing of eggs cannot be too heavily emphasised.

Growth and development of the foetus are not considered within the scope of this volume, although some final maturational events in the foetus are discussed in the next chapter. Gestation lengths have been presented in Table 3.6, and there is good evidence for breed effects in this regard. Twin calf or multiple lamb foetuses are usually associated with shorter pregnancies.

Parturition: perinatal and post-partum problems

4

The reproductive cycle is brought to fruition with successful delivery of one or more viable offspring. It is salutary to note, however, that despite the investment of time and capital in sustaining a pregnancy, there is a relatively high death rate of mature foetuses just before, during or just after the birth process. Research leading to a greater understanding of the changes in mother and foetus at term could be justified on the grounds of such losses alone, because further information on the physiology of birth should enable development of treatments to facilitate the process, or indeed to control it quite precisely (see Ch. 6, p. 130). The subsequent interval until the return to fertility is also one of economic significance, and the changes occurring in the post-partum animal need to be described if the reproductive cycle is to be fully understood.

Preparations for birth: foetus and mother

The mechanisms underlying termination of pregnancy after a gestation length characteristic for the species (Table 3.6, p. 68) are incompletely understood. However, due to coordinated growth of the placenta and foetus together with accumulation of placental fluids, the uterus becomes massively distended as term approaches. Foetal maturity is expressed in a number of ways, such as limb movements, swallowing movements, and activity of the trachea, lungs and diaphragm. Eventually a situation is reached in which the foetus is in considerable physiological stress. A relevant observation here is that while functional maturation of many organ systems must occur in the last weeks of gestation, growth of the foetal adrenal glands is particularly prominent (Fig. 4.1), and these glands play a major role in precipitating birth. It is reasonable to suppose that a prolonged existence *in utero* would represent a greater stress than presentation to the rigours of the world outside.

The specific components of stress registered by the foetus in late gestation are also unknown, but presumably involve (1) physical, (2) respiratory and (3) more general metabolic stresses. The first of these – the limitations of space – has already been referred to, while respiratory stress may be brought about by a combination of inadequate gaseous transfer across the placenta and as a response to foetal lung movements. At a more general metabolic level, the movement of nutrients and waste products across the placenta may no longer be performed effectively, leading to a further form of stress. Other indices of foetal maturity doubtless come into play, for an event of such overwhelming significance as the termination of gestation is likely to have a multifactorial basis. In any event, we can state on the basis of sound experimental evidence that just as the embryo establishes pregnancy by imposing its influence on the mother, so the foetus is primarily responsible for terminating pregnancy. How is this brought about and what steps are involved?

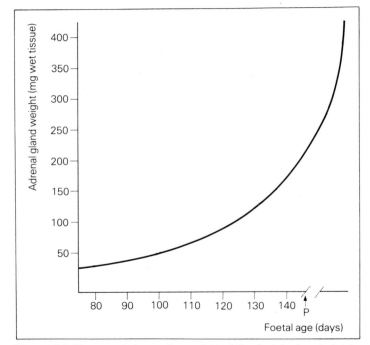

Figure 4.1 The steeply increasing weight of the foetal adrenal glands shortly before the onset of parturition in sheep. The adrenal cortex plays a key role in mediating the endocrine changes that lead to termination of pregnancy

Role of progesterone

Maintenance of pregnancy requires continued secretion by the mother of the steroid hormone, progesterone, the principal source being the corpus luteum (or corpora lutea). Secretion of progesterone by these structures is supplemented in species such as the sheep by placental progesterone but, in the absence of the ovaries, sufficient hormone becomes available to maintain a gravid uterus only after approximately day 50 of gestation. Progesterone is referred to as the 'hormone of pregnancy' although, of course, the whole spectrum of maternal endocrine activity is normally involved, and termination of pregnancy therefore requires some means of curtailing secretion of progesterone. This is not least since the hormone has acted to suppress contractile activity of the uterus during gestation (i.e. the so-called 'progesterone block'), whereas sensitisation and excitability of the uterine muscle are necessary preliminaries to expulsion of the foetus. In addition to changes in the uterus itself, the cervical canal has to be prepared for dilation by the foetus, and the pubic ligaments need to become relaxed for the final stages of delivery. An active mammary gland must also be available to the newborn and, although preparation of the gland has been proceeding throughout pregnancy, events in the last few days of gestation are critical to initiation of colostrum secretion and full lactation.

Hormonal switches for parturition

While the end of gestation involves functional interactions between mother and foetus for a successful delivery, the primary switch is thought to be in the foetal hypothalamus. In response to conditions of stress, the releasing hormone for adrenocorticotrophic hormone (ACTH) is secreted and this trophic hormone is produced in increasing amounts by the anterior pituitary gland of the foetus. This, in turn, causes enhanced secretion of corticosteroids into the foetal circulation, and it is the latter step that underlies cessation of progesterone secretion. Thus, a cascade of *foetal* endocrine changes acting through the placenta prompts removal of the *maternal* block to uterine muscular activity, and overcomes maintenance of pregnancy.

The key role of the foetal hypothalamic-adrenal axis has been demonstrated experimentally in several ways (Fig. 4.2). Removal of the foetal pituitary or adrenal glands prolongs gestation, whereas infusion of ACTH or cortisol into the foetal circulation near term initiates parturition. Moreover, foetuses with spontaneous congenital malformations of the head or hypoplasia of the adrenal cortex undergo much prolonged pregnancies. A further point concerning the function of the foeto-adrenal cortex near term is that cortisol secretion promotes maturation of enzyme systems in organs as diverse as the lung, liver and thyroid, and is specifically involved in surfactant deposition in the alveoli of the lungs.

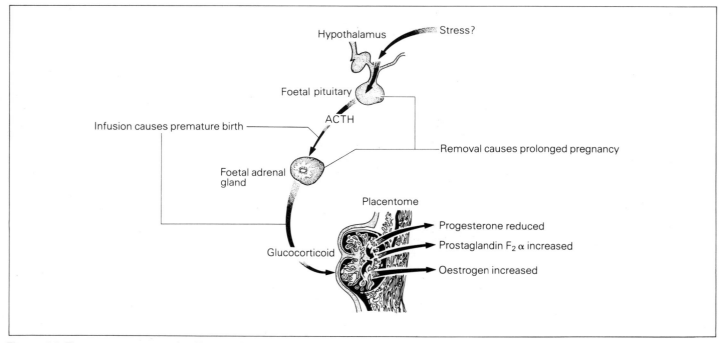

Figure 4.2 The sequence or cascade of hormonal steps associated with the termination of pregnancy in sheep. Foetal stress in very late gestation leads to enhanced secretion of foetal glucocorticoids which, directly or indirectly, cause a reduction in the circulating concentration of progesterone and an increase in the sensitivity and contractile activity of the uterus

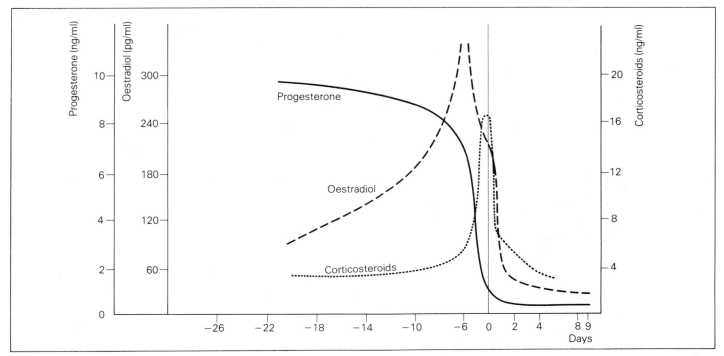

Figure 4.3 Endocrine changes in late gestation in the cow showing the relationship between decreasing progesterone, increasing oestrogen, and the peak of corticosteroid activity in maternal blood. Corticosteroid activity in the foetal circulation would have been more conspicuous

This protein is essential in the preparation of a physiologically functional offspring, as the absence of surfactant would lead to collapse of the lungs due to surface tension effects. Regulation of glycogen reserves is also influenced by foetal cortisol secretion, mobilisation of carbohydrate stores being essential at the time of birth.

Increased secretion of cortisol by the foetus during the last few days of gestation leads to a steep decline in the concentration of progesterone circulating in the mother (Fig. 4.3). This decline is achieved in part through the mediation of prostaglandin $F_{2\alpha}$ ($PGF_{2\alpha}$) secreted by the uterine wall in pigs or more specifically by the placentome (the unit of placental attachment) in ruminants. As already noted when discussing the oestrous cycle in the first chapter, $PGF_{2\alpha}$ is the luteolytic hormone that brings about cessation of progesterone secretion by the corpus luteum. It is the luteolytic property of $PGF_{2\alpha}$ that also causes regression of the corpora lutea of pregnancy in pigs, cows and goats, whereas in sheep the situation is more complex. Here the placenta is a major source of progesterone synthesis in late gestation and while $PGF_{2\alpha}$ causes regression of the corpus luteum, placental secretion of progesterone has to be curbed under the influence of foetal cortisol. This is achieved by enzymatic conversion of progesterone to oestrogen. In one sequence in sheep, therefore, the progesterone block is removed and the uterus is sensitised by oestrogen to resume contractile activity under the influence of prostaglandins. The foeto-placental unit is also the source of the oestrogens in pigs and cows that sensitise the uterine muscle, and these oestrogens enhance the release of prostaglandins at the time of labour.

Related physiology and delivery of the foetus

Before actual delivery of the foetus begins, it becomes orientated in the uterus in a manner enabling the easiest passage through the pelvic girdle and birth canal. In species such as cows and sheep, the foetus may have been on its back for much of gestation but it rotates shortly before birth almost to an upright position (Fig. 4.4). The nose and front legs become directed towards the cervix, and the conventional anterior presentation involves the foetus emerging front feet first. Posterior or breech presentation of the foetus in cows and sheep is usually associated with difficulties of birth (dystocia) whereas in pigs, by contrast, the foetuses do not turn around during late gestation and anterior and posterior presentation seem equally common (Fig. 4.5).

The actual process of delivery can be regarded in three stages. First, that of uterine contractions forcing the fluid-filled foetal membranes against the internal os of the cervix, leading to dilation in a matter of hours. Second, emergence of the foetus with rupture of the foetal membranes. Contraction of the abdominal muscles (straining) is conspicuous during this stage. Third, delivery of those

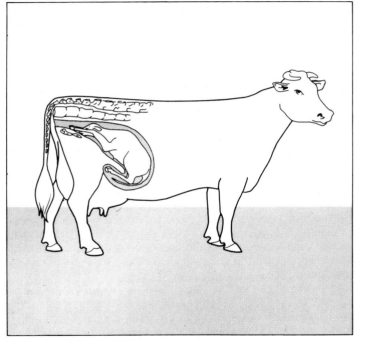

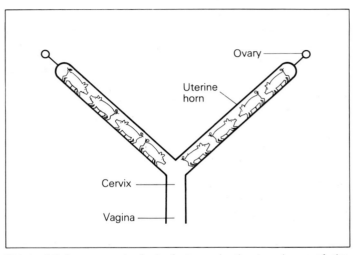

Figure 4.5 Arrangement of pig foetuses in the two horns of the bicornuate uterus to indicate that there is no particular form of orientation shortly before birth. Anterior and posterior deliveries appear equally common

Figure 4.4 Orientation of the foetal calf in late gestation, and especially shortly before the onset of parturition. This form of presentation to the cervical canal enables the easiest form of delivery once the uterine and abdominal contractions of labour begin

placental membranes that do not surround the foetus. Together these are termed the three stages of labour.

Once the head of the foetus engages the birth canal, this stretches the cervical muscle which causes a reflex release of oxytocin from the posterior pituitary gland of the mother. This release of oxytocin in conjunction with uterine prostaglandins causes the powerful contractions characteristic of labour and propels the foetus through the cervical canal. The polypeptide hormone, relaxin, from the maternal ovaries and placenta has already acted to soften the tissues of the cervix and increase their distensibility, and a further specialisation is the secretion of mucus to facilitate passage of the foetus. Delivery itself involves rupture of the umbilical cord, liberation of the foetus from retained placental membranes, and stimulation in the foetus of an independent cardiac and respiratory physiology. The *foramen ovale* and *ductus arteriosus* of the newborn's heart close so that its pulmonary circulation becomes effective, and this clearly relies on active breathing movements in the newborn; the latter are usually seen as a reflex (a) after rupture of the umbilical cord, (b) after release from the constrictions of the birth canal, and (c) in response to the cooler ambient temperature compared with that of the uterus.

Assuming that breathing and a functional blood circulation are assured, and that the foetus is not at risk due to conditions of hypothermia, then it can exist for a period of hours by mobilising reserves of muscle and liver glycogen. However, initiation of suckling is urgent, not only as the source of nutritional support but also for the passive immunological protection conferred on the newborn by antibodies present in the colostrum. Moreover, while colostrum may be available from the dam for 24–48 hours or more, the immunoglobulin antibodies are only absorbed efficiently across the intestinal wall during the first few hours post partum.

Separation of placenta

Much of the placenta appears as the afterbirth during the third stage of labour. The epithelio-chorial placenta of the pig permits a straightforward separation of the allanto-chorion from the uterine surface, and retention of the placenta is seldom a problem. In ruminants, by contrast, interdigitation of the maternal and foetal components of the placenta occurs within each placentome (Fig. 3.16, p. 66), and delayed delivery or indeed retention of the afterbirth is more common in ruminant species. This is especially the case in cattle with twins or in sheep with multiple offspring, and the explanation is almost certainly that the more advanced foetus triggers birth of both or all foetuses before separation of the placental membranes has been completed. Accordingly, retention of afterbirth material in such instances is associated with a less developed foetus.

After expulsion of the placenta(e), contractions of the uterus continue for some time and serve to diminish bleeding where tearing of the uterine epithelium is extensive. Further phases of contraction will occur in response to the oxytocin that is reflexly released during each episode of suckling. Closure and restoration of the cervical tissues also follows completion of birth, and prevents bacteria gaining access to the uterus (an invasion would lead to infection and development of pus – pyometritis). Even so, trauma due to laceration of the cervical canal may have occurred during delivery, and substantial modifications may be required in this tissue.

Birthweights and sex ratios

Liveweights of the foetus at birth are presented for cattle, sheep and pigs in Table 4.1. Males tend to be heavier than

Table 4.1 Some common values for foetal liveweights for animals kept under Western conditions of husbandry

Species	Range (kg)	Mean (kg)
Calf	27–50	36*
Lamb	3.2–4.9	4.1*
Piglet	0.9–1.5	1.2

*Lower for twins or triplets.
Substantial breed differences in cattle.

females, and the heavier newborn – if its delivery is not unduly difficult – will be stronger, more vigorous, and less likely to succumb to hazards in the first hours or days after birth. In litter-bearing species such as pigs, the location of the foetus in the uterus relative to litter mates and major uterine blood vessels was thought to have an important influence on birthweight. However, extensive studies involving identification of foetuses *in utero* by tattoo numbers have failed to support this notion.

It is known that appropriate regimes of feeding in the last portion of pregnancy can increase foetal weight and thereby potential viability, but against this apparent advantage are offset the problems of birth of overlarge foetuses and the possible costs of veterinary intervention. Moreover, difficult calvings or farrowings are frequently associated with a delayed return to the fertile state, and in cattle particularly there may be sound reasons for monitoring the maturity of the calf in late gestation (see below).

The sex ratio when examined in a sufficiently large population of newborn is very close to 50 : 50, with perhaps a tendency to male predominance (Table 4.2). Deviations from a 50 : 50 sex ratio are reported periodically, but these are

Table 4.2 Sex ratios at birth (secondary sex ratios) in large sample populations of cattle, sheep and pigs to indicate the proximity to 50 : 50 (*NB*. High *male* and low *male*) ratios

Species	Population sampled (no.)	Ratio of males (%)	Population sampled (no.)	Ratio of males (%)
Cow	4 950	51.8	985	48.6
Sheep	50 700	49.5	8 950	49.2
Pig	2 350	52.8	16 250	48.8

minor aberrations in comparatively small numbers of animals, and are seldom repeatable or statistically significant. In fact, the sex ratio at fertilisation in mammals (the primary sex ratio) is invariably in favour of males, which infers a relatively greater loss of male embryos and/or foetuses as gestation proceeds. If the basis of this differential loss were understood, it might become possible to accentuate and accelerate the trend so that a higher proportion of females appeared at term.

Perinatal loss

Perinatal loss refers to the death of offspring immediately before, during or shortly after birth. In pigs, this leads to a reduction in litter size whereas in cattle, and to a lesser extent in sheep, perinatal loss means both a completely unfruitful pregnancy and loss of time and overheads invested in that pregnancy. The extent of this loss in actual numbers of animals and in economic terms is enormous. Because much of it is accessible in a way that prenatal loss is not, a continued acceptance in the 1980s of levels of perinatal loss as high as 15–20 per cent or more is a sobering comment on British farming practice. When we compare the situation in farm animals with that which now obtains for childbirth in Britain (1.5% mortality and even less in Scandinavia and North America), the conclusion is unavoidable that perinatal loss could be very significantly reduced by sufficient inputs of

specialist attention. It is therefore pertinent to examine the factors contributing to perinatal loss and to ask whether the continuing wastage of fully formed offspring is justifiable.

Major causes of perinatal mortality are listed in Table 4.3. The first of these – conditions of toxaemia in the dam in the last days of gestation – should be avoidable by appropriate husbandry, especially with regard to nutrition and/or exercise. As to problems associated with actual delivery of the foetus, those of premature placental separation or premature tearing or twisting of the umbilical cord leading to asphyxiation are not common. On the other hand, incorrect presentation of the foetus and/or a protracted delivery due to uterine inertia are frequent occurrences, and require skilled intervention to save the offspring and prevent the mother being at risk. While orientation of piglets at birth does not appear critical, the fact remains that about 7 per cent of all normal, fully formed piglets that were alive in the uterus just before birth are stillborn. Of the piglets born alive, a further 13 per cent die within 48 hours of birth, thereby giving a perinatal loss of about 20 per cent of the pig crop. Expressed even more dramatically, over 50 per cent of the pre-weaning deaths of piglets occur before the end of the second day of life. Because only a few piglets are affected in each litter, scientific impetus to overcome the problem has been insufficient.

Of the further causes listed in Table 4.3, most could be prevented by provision of suitable facilities, attention to hygiene, and sufficient inputs of skilled labour or a veterinary

Table 4.3 Some causes of perinatal loss in farm animals and suggested procedures for their remedy

Condition	Recommendation
Metabolic disturbance of mother before or during birth (e.g. pregnancy toxaemia)	Appropriate feeding and husbandry
Protracted birth, placental subfunction and/or asphyxia	Veterinary assistance
Abnormal presentation during labour; dystocia	
Hypothermia (environmental chilling)	Intra-peritoneal injection of glucose solution
Failure to suckle or agalactia (lack of milk)	Stomach tube with colostrum
Crushing by mother, especially of piglets	Suitable accommodation and heat lamps
Low birth weight and weakness	No simple remedy, apart from sound feeding in late gestation
Bacterial infection, scouring, dehydration	Attention to hygiene, and provision of antibiotics and mineral supplements
Anaemia, hypoglycaemia, mineral deficiencies	
Congenital and genetic abnormalities	Avoid intensive inbreeding
Mismothering, rejection or predation, especially of lambs	Supervision and use of lambing pens

surgeon. Of course, a small proportion of offspring will be stunted (i.e. runts) or congenitally malformed or undesirable on other grounds but, putting these on one side, it is not easy to accept that levels of perinatal mortality should not be almost halved – albeit at a price. The argument is the more forceful in situations where farmers are trying to increase the fecundity of their animals by techniques involving nutritional flushing, hormonal treatments or even transplantation of embryos: intrinsic fertility of the animals in question may already be sufficiently good, but perinatal losses need to be reduced. At the end of the day, however, sensitive economic judgements will be necessary before investing in some form of obstetrical clinic for farm animals.

The problem of low birthweight is more complex. While appropriate nutrition of cows and ewes should enable the foetus to attain a viable mature weight, there will invariably be some small foetuses within the litter in pigs. Even though the importance of birthweight in relation to piglet survival has been highlighted (Table 4.4), there remains the troublesome fact that a uniform birthweight within litters is as critical as a high average birthweight in influencing survival. Cross-fostering could perhaps provide a partial solution, and be rather less expensive than the approach of artificial rearing.

If it is accepted that increased supervision of the dam at term would contribute significantly to a reduction in perinatal loss, then there may well be attractions in controlling the time of birth by an induction procedure (see Ch. 6, p. 130). As a

Table 4.4 The relationship between birthweight and mortality in piglets in the interval until weaning at 8 weeks post partum (adapted from English and Smith, *Vet. Ann.*. 1980)

	Birthweight (g)		
	<907	907–1 361	>1 361
Total number of livebirths	130	440	399
Number of livebirths dying by 56 days	97	95	44
Percentage of livebirths dying by 56 days	74.6	21.6	11.0

majority of animals will usually give birth during the night or in the early morning, a treatment that accurately controls the onset of labour (without undesirable side effects) should be especially welcome to the stockman.

Lactation; and lactational and seasonal anoestrus

As already noted, endocrine changes underlying the events of parturition also programme the mammary gland for active lactation. A necessary synchrony is therefore found between arrival of the newborn requiring nutritional support and secretion of this support. The primary programming of milk synthesis and secretion in the mammary glands is endocrine through the complex of hormones termed the *lactogenic complex* (e.g. the pituitary hormones prolactin, growth hormone, thyrotrophic hormone and ACTH; insulin; and the steroid hormones oestrogen and progesterone), but

production of milk is also regulated by the extent of the suckling stimulus and the concomitant removal of milk from the udder. The duration of lactation in terms of weeks or months depends on many factors, again most notably on the frequency and magnitude of the suckling stimulus which, in the biological situation, will reflect the growth and capacity for independent feeding of the offspring.

While the mother is in the active phase of lactation, hormones of the pituitary gland are directed more towards supporting this process than to stimulating ovarian function. Thus, resumption of growth of ovarian follicles, their secretion of oestrogens, and the return of sexual behaviour characteristic of oestrus, together with ovulation, are generally held in abeyance during the first weeks of lactation. This would seem to represent a form of biological economy on the part of the dam whereby the demands of a further pregnancy are not imposed concurrently with the demands and stresses of the heaviest phase of lactation. A key physiological explanation of this situation is that there is an inverse relationship in the first weeks post partum between the anterior pituitary secretion of prolactin on the one hand and the gonadotrophic complex on the other. Thus, prolactin is actively secreted to support milk synthesis at a time when there is only minimal secretion of the gonadotrophins FSH and LH (Fig. 4.6). As the suckling stimulus diminishes with the interval post partum, so the secretion of gonadotrophic hormones is able to increase and results in growth of ovarian

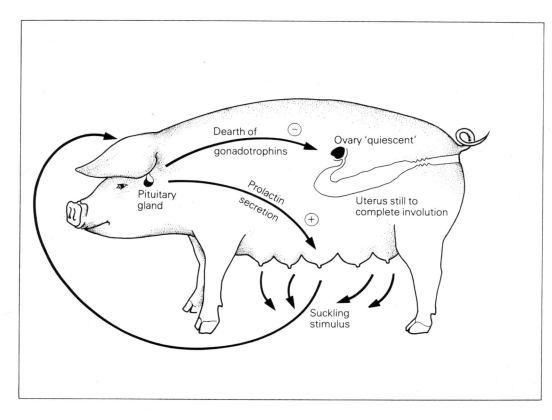

Figure 4.6 Endocrine relationships in the *post-partum* sow showing the active secretion of prolactin early in the suckling period, and the relative dearth of gonadotrophin secretion associated with lactational anoestrus and incomplete involution of the uterus. A change in the relative secretion of gonadotrophic hormones and prolactin occurs as the incidence of suckling decreases with the interval after birth

follicles. Precisely how these changes are achieved is still unknown, but the physiological status of the animal is monitored centrally and inhibition in the hypothalamus of the releasing hormone for gonadotrophin is gradually lifted while a lesser suckling stimulus will itself lead to lowered secretion of prolactin.

In general terms, the situation in the post partum animal is therefore a phase of active lactation in conjunction with one of minimum ovarian activity. Behavioural oestrus is not usually found in cattle in the first few weeks after calving (although silent ovulations may occur), nor is it seen in this interval in breeds of sheep not subjected to seasonal limitations of breeding (see below). This phase is therefore referred to as one of *lactational anoestrus* when oestrous cycles, ovulation and the opportunity to initiate a further pregnancy are withheld. Dairy cows generally return to oestrus 4–6 weeks post partum, but beef cows suckling their calves take 3–4 weeks longer. A low energy intake leads to protracted anoestrus. Pigs also undergo several weeks of lactational anoestrus in the sense of an absence of ovarian activity (and hence fertility), although this statement is not always accepted since sows commonly show oestrus 3–7 days after farrowing when they will certainly stand for the boar. However, this oestrus does not reflect mature ovarian follicles: it is anovulatory and is caused by the surge of foeto-placental oestrogens released at term (see above). As a purely experimental demonstration of this fact, the sow's ovaries can be removed just before birth of the litter and oestrus will still be observed a few days later.

Since lactational anoestrus may last for 4–8 weeks or more in conventional Western systems of animal production, the question naturally arises as to whether there are simple and straightforward procedures for reducing its duration. In terms of the breeding cycle, lactational anoestrus can be regarded as an unproductive phase, and its duration has a major influence on the number of offspring or litters produced per unit time. Weaning the offspring at, or a few days after, birth is one approach to overcoming lactational anoestrus, but because specialist facilities for feeding the prematurely weaned animals are required, this becomes expensive and involves added inputs of labour. This is well illustrated by the stainless steel decks of heated battery cages used for very early weaned piglets – a costly and inflexible investment that is unlikely to appeal to most pig producers. However, the physiological point here is that by removing the suckling stimulus from the sow, there is a rebound of gonadotrophin and, thus, ovarian activity as prolactin secretion rapidly diminishes; cycles are resumed within one to two weeks. A simpler and less drastic approach is to introduce creep feed for the young and to reduce the frequency of suckling, thereby accelerating the resumption of oestrous cycles in the mother. This is almost certainly aided by the introduction of a mature male, principally due to the influence of his odours (pheromones).

Further approaches involve hormonal treatments, and

these are less likely to be successful unless imposed with care and understanding, although it is possible to treat cows with progestagens administered in intra-vaginal sponges or coils for a period of 2–3 weeks to blockade pituitary activity, after which the device is removed and there is a rebound of ovarian follicular growth (see Ch. 6, p. 105). If such a treatment is imposed after the first 3 weeks of lactation, it is generally successful in inducing oestrus and ovulation. Other forms of hormone treatment involving injection of gonadotrophins are available for pigs, but these can no longer be recommended as more consistent results are obtained simply by 3- or 5-week weaning (Table 4.5).

Table 4.5 The influence of the time of weaning on the number of litters per year and the number of piglets weaned

Criterion	3-week weaning	5-week weaning	8-week weaning
Days per cycle	148	162	183
Litters per year	2.5	2.2	2.0
Piglets weaned	25	22	20

Seasonal anoestrus

Most breeds of sheep show seasonal reproductive activity, responding to the changing pattern of daylength (photoperiod) by coming into oestrus and breeding in the autumn as daylength decreases. Accordingly, they lamb in the spring and are then exposed to increasing daylength, a situation in which there is little ovarian activity due to negligible secretion of gonadotrophic hormones. The situation of lactational anoestrus in sheep, therefore, is usually masked by that of seasonal anoestrus, which may last for 4–5 months. No such phenomenon is found in cattle and pigs maintained under reasonable conditions of husbandry.

Because seasonal anoestrus limits the overall productivity of the flock, several different avenues have been explored to overcome this constraint. Breeds such as the Dorset Horn and Finnish Landrace show only a restricted phase of seasonal anoestrus, so if it were possible to cross with these breeds, or indeed to introduce the purebreed, this might prove advantageous. An alternative approach, but much less convenient, would involve housing ewes in lightproof buildings and subjecting them to a regime of decreasing daylength to mimic the pattern of autumn. This is clearly expensive from a number of points of view, and is unlikely to have widespread commercial attraction. Perhaps the most straightforward approach of all is to treat the ewes with intra-vaginal sponges containing progestagens for 14–16 days and rely on the gonadotrophin rebound effect upon their withdrawal. This method and various modifications are described in Chapter 6. The only general point to make here is that such treatments do not work when ewes are in deep anoestrus; their success in eliciting a fertile response improves as the breeding season approaches.

Involution of uterus

While the constraints of lactational and seasonal anoestrus have focused attention on lack of ovarian activity, a further limitation to a prompt post-partum restoration of fertility concerns the condition of the uterus. After the distortion and massive distension of this organ during gestation – not only by the foetus but also by the placental membranes and associated fluids – the uterus needs to undergo a phase of shrinkage, repair and glandular regeneration. The period required to restore the uterus to a condition in which it could support sperm transport and embryonic development is referred to as the period of *involution* (see Fig. 4.6, p. 85). Before getting into detail and possibly losing perspective, it should be noted that uterine involution is invariably completed before the animal escapes from the period of lactational anoestrus; indeed, there may be a functional relationship between the two. So although involution is presented as one of the limitations to re-establishment of pregnancy, in reality this will prove to be so only under regimes of early weaning.

A major factor promoting the first stages of involution is the influence of oxytocin, released by the suckling reflex, on the smooth muscle layers causing repeated contraction of the organ, decrease in length and volume, and removal of any remaining blood and placental debris. A further gross change is the loss of water from the uterine tissues, as can be seen in the rapid decrease in weight (Fig. 4.7). Subsequent

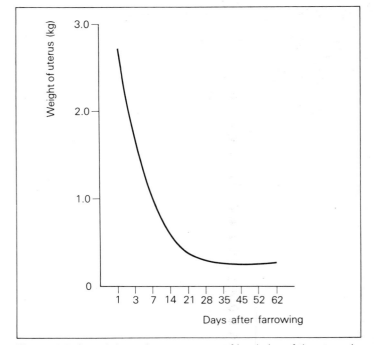

Figure 4.7 One of the major components of involution of the uterus is the progressive loss in weight, here illustrated in the first few weeks after farrowing

modifications concentrate on the connective tissue and epithelial layers. In pigs, epithelial modification is relatively simple in terms of regenerating cell layers that will support gamete and embryo survival, but in cattle and sheep there is the additional requirement of modifying the caruncles – the specialised maternal components of the placentome. These undergo progressive shrinkage, followed by necrosis and sloughing of cells from the vestigial glands, before elaboration of a new epithelium. Sloughing and liquefaction of the maternal tissues produce the persistent lochial discharge normally seen in cattle for 10–12 days post partum. The potential caruncles of the next pregnancy are barely visible, and do not interfere with transport and storage of spermatozoa.

The time required for involution of the uterus in farm animals is presented in Table 4.6, and suggests a period of weeks rather than days. Even so, the degree of involution is pronounced in the first seven or eight days following birth, but then proceeds more slowly. It should be stressed that the figures in Table 4.6 are estimates based on morphological and histological criteria, and that if the extent of uterine involution were tested in a functional sense by the transplantation of embryos, then rather shorter intervals might be recorded. One reason for this suggestion is that a number of laboratory animals (e.g. guinea-pig) have a fertile post-partum oestrus in which ovulation and fertilisation can occur within hours of delivery. So, even though the manner of implantation is different in rodents, an endocrine balance appropriate for ovulation may also lead to suitable conditions in the uterus. A second reason is that attachment of the embryo to the uterine epithelium does not begin for some 2–3 weeks after mating in farm animals (see Table 3.5, p. 61). Until this time, the embryo is free-living in the lumen and not dependent on structural interactions with the maternal epithelium.

Table 4.6 The expected time course of uterine involution in farm animals

Species	Details	Time required (days)
Cow	Initial shrinkage of uterus within 7–10 days of birth Caruncles sloughed and discharge completed by 10–12 days New epithelium formed by 25–30 days Uterus fully restored by 40–45 days	35–40
Sheep	Necrosis and sloughing of caruncles completed within 7–8 days Regeneration of epithelium within 25–30 days Uterus restored to cyclic dimensions in 30–35 days*	25–30
Pig	Some discharge for up to one week post partum No sloughing of endometrial tissue as in ruminants Normal columnar epithelium formed by 21 days Uterus fully restored in size by 21–28 days	25–28

*May be delayed by seasonal anoestrus.

Therefore, reiterating a point already made, if hormonal conditions are appropriate in post-partum farm animals for a return to oestrus and ovulation, then involution of the uterus should have occurred. However, it still remains to be shown conclusively that the fluid environment of the uterine lumen following very early weaning and rebreeding is competent to support development of a full number of blastocysts, and that qualitative factors such as specific proteins are not limiting.

Restoration of fertility

After the constraints on reproduction of lactational and/or seasonal anoestrus, and when involution of the uterus is completed, potential fertility is signalled once more by regular oestrous cycles and receptivity to the male. Nonetheless, there is good evidence – especially for cattle – that conception rates increase progressively with the first few cycles after the period of anoestrus (Table 4.7). Moreover, in situations where early weaning of pigs is practised, the number of eggs ovulated seems also to increase when one or more cycles are allowed to elapse before rebreeding the sow. This comment on ovulation rate is at present inference, for there are certainly other possible explanations for the generally smaller litter size the earlier post partum that the sow is rebred.

In the light of the above observations, farmers may well choose to allow animals to have one or more post-partum

Table 4.7 The increase in conception rate in dairy cattle with the number of oestrous cycles elapsing since parturition

Oestrous period at first insemination	Animals inseminated (no.)	Animals conceiving (%)
First	17	35.3
Second	28	50.0
Third	67	73.1
Fourth	61	72.1
Fifth	7	100.0

cycles before breeding in the expectation of an improved conception rate and/or litter size. Balanced economic judgements are again involved here, since the cost of maintaining animals in a barren condition during each period of three weeks could be quite significant in relation to potential return.

Subfertility, infertility, sterility

5

There is an important distinction between infertility and sterility: infertility is a temporary condition in the sense that it is susceptible to treatment, whereas sterility is invariably permanent. Sterile males, for example, may have testes devoid of germ cells and no amount of hormone treatment will be able to initiate sperm production in such animals. Sterile females commonly have congenital abnormalities of the reproductive tract, such as missing oviducts or blind uterine horns. Infertility, by contrast, may be a response to some form of stress, such as the demands of a heavy lactation or the overcrowding of animals in indoor pens. Alternatively, infertility may be associated with nutritional constraints, such as a shortage of specific minerals or vitamins, or again it may be due to infection. The essential point is that infertility may be overcome spontaneously or it is amenable to some form of treatment. Maintenance of adequate records for individual animals and close cooperation with a veterinary surgeon are the usual prerequisites for restoring the animal to fertility.

Subfertility is frequently used as a clinical term to refer to situations in which the fertility of a herd or flock falls significantly below its customary level. The underlying cause may be a particularly rigorous winter acting in concert with nutritional deficiencies, or infectious disease may be present in the herd. Discussion of the role of infectious diseases is beyond the scope of this volume.

Decline in fertility with age

Although litter size in many species of mammal tends to increase with the interval elapsing after puberty, fertility eventually declines with age after a certain number of pregnancies. In other words, if farm animals are maintained in the breeding herd for a sufficient period of time, then litter size commences to decrease or calving interval to increase. The phenomenon of diminished litter size in pigs due to age is seldom witnessed under Western systems of production since animals tend to be culled before this stage is prominent. However, diminishing litter size can be demonstrated in a number of laboratory species, and experimental evidence derived from these species indicates that the limitations to reproductive performance are associated primarily with defects in the uterus rather than in the eggs or ovaries.

By means of reciprocal embryo transplantation studies between 'young' and 'old' mothers in such species as mice, hamsters and rabbits, it is possible to demonstrate that embryos from old mothers yield litters of normal size when transplanted to the uteri of young mothers. By contrast, embryos derived from young mothers yield smaller litters when transplanted to the uterus of ageing females. Moreover, smaller litter size in ageing females is due to increased embryonic death occurring *after* implantation – a finding which implicates the condition of the uterus rather than that of the embryos. The trauma of previous pregnancies may have influenced the uterus in ageing females, and histological examination shows accumulation of scar tissue and of fibrous connective tissue together with a diminished vascularisation of the uterine stroma. These degenerative conditions clearly influence the potential for placental development. Even though the manner of implantation differs between laboratory rodents and farm animals, there is little doubt that an ageing uterus would act to limit fertility or litter size if farm animals were kept for a sufficient number of pregnancies.

Bearing in mind that the stock of oocytes is formed in the ovaries by the time of birth and that there is no replenishment thereafter (see Ch. 1, p. 7), ovulation of potentially defective oocytes might also be a factor contributing to diminished reproductive performance with advanced parity. Minor but progressive degeneration in the structure and viability of oocytes with the interval after birth would seem a reasonable possibility, but this is not an easy topic to study experimentally for the following reason. Even though increasing numbers of degenerating oocytes may be demonstrated *in the ovaries* with increasing maternal age, this does not mean that those oocytes *released at ovulation* are necessarily lacking in viability. In other words, the eggs that are ovulated may represent a selected population. Moreover, in a large-scale study of ovarian histology in ageing cows, significant differences could not be demonstrated between fertile and infertile animals in the numbers of healthy ovarian follicles and eggs.

Although cattle may remain capable of breeding for 15 years or more, the calving interval tends to increase after five or six pregnancies, usually leading in practice to elimination of such animals from the herd. The cause of an increasing calving interval is open to several interpretations, but failure of fertilisation, abnormal fertilisation or death of the embryo during the pre-attachment stage probably explain the majority of instances of what is termed 'repeat breeding' (Table 5.1). Evidence obtained from extensive programmes

Table 5.1 The proportion of cattle with reproductive problems causing return to service (repeat breeding)

Nature of problem	First service: heifer (%)	Repeat breeder heifer (%)	cow (%)
Abnormal reproductive organs	2.7	13.5	6.0
Failure of fertilisation	21.7	35.3	39.3
Death of embryo within 30 days	16.0	24.8	32.5
Morphologically normal embryos at 30 days	59.6	26.4	22.2

of artificial insemination in dairy cows indicates that fertility increases slightly up to 3–4 years of age – largely due to culling of heifers with anatomical or endocrine abnormalities – and then gradually declines in cows of 6–7 years or older. In beef cattle, by contrast, fertility may not decline until 9–10 years.

From a practical point of view, it could be argued that the phase of reproductive decline is not a serious constraint as long as its onset can be recognised and suitable replacement animals are available to permit a policy of culling. Nonetheless, the situation remains unsatisfactory from a scientific point of view, since an understanding of the precise causes of reproductive decline and its chronology might enable some form of therapy to be instigated. In this manner, animals of high merit for traits such as milk production in cattle or mothering ability in sows might be retained in the breeding herd for several extra pregnancies at very little extra cost.

Information on reproductive decline in sheep is less easy to come by than in cattle because of the extensive conditions under which sheep are usually bred, with the rams in attendance for the duration of several oestrous cycles. However, evidence from flocks maintained at research stations and on experimental farms suggests that maximum birth rate is reached at approximately 5–6 years of age, after which there is a gradual decline (Fig. 5.1). In pigs, it is generally accepted that there is an increase in litter size for the first 4–5 pregnancies and then a gradual decrease. Increasing litter size is associated with age rather than parity (i.e. the number of litters), but the reasons for the eventual decline in reproductive performance in sows have yet to be resolved. Since the number of eggs ovulated continues to increase with age (Table 5.2), some of the eggs shed by ageing sows may not

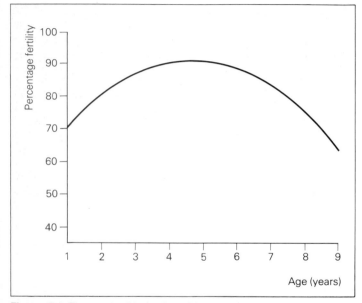

Figure 5.1 The curve of fecundity with respect to age of the dam. This increases for the first 4 or 5 pregnancies in sheep, and then gradually diminishes. Approximately similar curves can be plotted for pigs and cattle

Table 5.2 To illustrate the increase in number of eggs ovulated (ovulation rate) with increasing age in pigs

Criterion	Number of previous litters							
	0 (gilts)	1	2	3	4	5	6	7
Number of animals	35	20	20	17	12	6	8	16
Mean ovulation rate	13.1	17.9	18.8	21.8	21.9	19.8	22.6	24.4

be fully viable or there is a deleterious influence on embryonic survival of the ageing uterus, or both these factors may be operating simultaneously.

Whatever the causes of declining fertility in individual farm animals, a sound system of recording is essential before the extent of the decline (e.g. increasing calving interval) can be appreciated and an appropriate culling policy decided upon.

Ovarian disorders: cysts, persistent corpora lutea, short cycles

Ovarian disorders associated with temporary or prolonged periods of infertility include follicular and luteal cysts in one or both ovaries, and the quite common condition of a persistent corpus luteum. Oestrous cycles are disturbed or in abeyance in the face of these disorders.

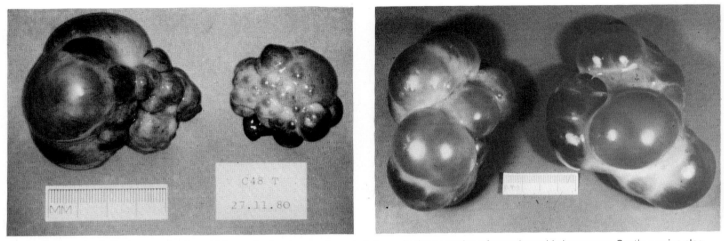

Figure 5.2 Ovarian follicular cysts in pigs which usually develop due to an imbalance in the secretion of gonadotrophic hormones. Cystic ovaries also arise as a sequel to most synchronisation treatments with synthetic progestagens. (Courtesy of Dr C. Polge)

Abnormal development of Graafian follicles, especially in the form of cysts (Fig. 5.2), leads to enlargement significantly beyond the ovulatory diameter. Follicular cysts may measure 3–8 cm in diameter in cattle and 2–4 cm or more in pigs, and in these the potential for ovulation is lost. Cystic follicles are thin-walled and show extensive cellular degeneration, especially in the innermost layers of cells – the granulosa.

Even so, the follicles can continue to enlarge and accumulate a clear fluid and they may remain active in secreting steroid hormones, especially oestrogens and androgens. These hormones in turn may cause anomalies of sexual behaviour such as the intense and persistent condition of nymphomania (with the associated bellowing noise), or prolonged periods of anoestrus. Another form of ovarian cyst is a luteinised follicle, in which several layers of luteal cells surround a fluid-filled or haemorrhagic cavity. In other words, these cysts appear part-follicle, part-corpus luteum. Luteal cysts are generally associated with anoestrus.

Both forms of cyst are thought to arise from an endocrine imbalance, often during the heaviest period of lactation, and both forms tend to persist and require clinical intervention rather than regressing spontaneously. Although the incidence of their formation may be influenced genetically (and the condition is prominent in certain inbred strains of pig), cystic ovarian problems also increase with age and therefore underlie instances of reproductive failure. Systemic injections of LH or synthetic gonadotrophin releasing hormone have been used successfully to treat cystic follicles in cattle, but the manner of response of a degenerating follicle to such hormone treatment is uncertain. The more widespread clinical approach to cysts in cattle is to palpate the ovaries *per rectum* and then to rupture the cyst by squeezing.

The expression of regular oestrous cycles will also be interrupted by the presence of a persistent corpus luteum. This condition is found in ageing cows and is commonly associated with infection of the uterus. Infection leads to the formation of a pus in the uterine lumen (pyometritis), and this in turn inhibits the normal functioning of the luteolytic mechanism (see Ch. 1, p. 14 and Ch. 3, p. 68). In brief, the uterus does not release the luteolytic hormone $PGF_{2\alpha}$ when there is pyometritis, so the corpus luteum remains active for weeks beyond its cyclic lifespan. Persistent corpora lutea are best enucleated (squeezed out) manually in cattle, after which the uterine infection may clear spontaneously as the animal returns to heat. Alternatively, antibiotic therapy to remove the pyometritis may result in the restoration of oestrous cycles, since the uterus will regain the ability to secrete $PGF_{2\alpha}$. The most recent clinical approach to overcoming persistent corpora lutea has been the intramuscular injection of a synthetic analogue of $PGF_{2\alpha}$ which causes regression of the luteal structure and permits follicular maturation.

A further abnormality concerning the corpus luteum is a short or inadequate luteal phase due to the structure not attaining full development. The condition is very difficult to predict or to analyse clinically except by blood sampling or hormone estimations, whereas it is revealed retrospectively in oestrous cycles significantly shorter than normal. Our current understanding is that a fully mature corpus luteum is not formed; instead the developing luteal body commences to break down within six or seven days of heat giving cycles of 9–12 days in length. Since the animal may have been mated or inseminated at this heat, luteal failure will result in embryonic failure. Once again, an endocrine imbalance is thought to precipitate this condition.

Male infertility: penile defects, cryptorchids and other factors

Although discussions on problems of fertility tend to focus on the female, there are specific disorders of the male associated with infertility or sterility which include penile deficiencies and cryptorchidism. Because the number of males associated with a breeding herd is small, male fertility or subfertility generally receives less attention than problems in the female, yet its influence can be widespread and of major economic importance. Short of a clinical investigation, however, diminished male fertility is usually revealed only through sound records in which returns to service or insemination are highlighted. In situations in which detection of oestrus – and thus returns to oestrus – may be less than optimal, some system of early pregnancy diagnosis (see Ch. 6, p. 124) may be a valuable means of monitoring male fertility.

Mature males normally show active sexual behaviour in the presence of oestrous females, leading to mounting and the pelvic thrusting characteristic of mating. Even so, intromission and ejaculation into the female tract cannot be taken for granted as a small proportion of males have abnormalities of the penis or associated musculature that deflect the male organ at the time of attempted intromission. The degree of sexual excitement may have been sufficient for ejaculation to have occurred and yet, because of the deflection, there would be no possibility of fertilisation and subsequent conception.

Sometimes the penis deviates to one side or, less commonly, ventrally so that the organ cannot be inserted during mating. Other forms of deviation are illustrated in Figure 5.3, and include openings of the urethra at the side rather than at the tip of the glans penis. While numerous surgical procedures have been described for correcting these abnormalities, elimination of such males from the breeding herd is usually recommended if suitable substitutes are available. As already indicated, the impact of an unsatisfactory male can be widespread and the economic pressures of modern farming scarcely permit us to gamble on the potential success of a doubtful sire.

In males kept on commercial farms, penile defects are normally revealed retrospectively during clinical investigations for suspected infertility, but in males used in artificial insemination and standing at collection centres, defects should have been seen during a preliminary veterinary examination. If not, they should certainly come to light during semen collection, be this by means of an artificial vagina or electro-ejaculation. In fact, with these artificial procedures, penile deviations should not prevent collection of the ejaculate, and this may be an attractive alternative for using genetically valuable males that would otherwise be despatched for slaughter.

Further penile problems may arise from damage sustained during mating – for example, if the female collapses or suddenly moves sideways. Damage caused in this manner

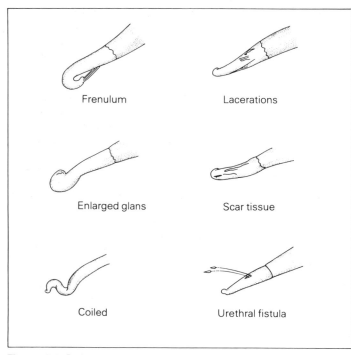

Frenulum

Lacerations

Enlarged glans

Scar tissue

Coiled

Urethral fistula

Figure 5.3 Defects or abnormalities of the penis in bulls which may be congenital or have arisen as a result of damage. Such defects may cause a diminished conception rate or serious infertility

usually involves tearing the connective tissue and rupture of blood vessels and, among farm species, is relatively common in the bull. Extensive swelling develops in the sheath anterior to the scrotum, and the haemorrhage will cause adhesions of the penis to the sheath and prepuce. Although the adhesions can be removed surgically, males that have suffered this form of damage rarely mate successfully again, either due to loss of confidence or to failure to extrude the penis fully. Yet another defect leading to a short term unwillingness to mate concerns lacerations of the penis; these may arise during unsuccessful mating or while trying to mount other males.

Cryptorchidism refers to the condition of undescended testes. Instead of leaving the body via the inguinal canals and entering the scrotal sac, the testes remain close to their original embryonic location or to the inner orifice of the inguinal canals and they are therefore exposed to abdominal temperature. This prevents formation of spermatozoa since the sperm-producing epithelium (the seminiferous epithelium) is destroyed at abdominal temperature and requires the cooler scrotal temperature to be functional. If the condition of cryptorchidism involves both testes, the animal is sterile. Even surgical withdrawal of the testes into the scrotum in such instances will be unable to initiate fertility since the stem cells (i.e. the spermatogonia) of the seminiferous epithelium will have been damaged or destroyed and cannot be replaced. If, on the other hand, the condition is unilateral and one of the testes has descended, the male should be fertile although sperm production will be significantly less than in a normal male. Not to be overlooked is the fact that male sex hormone production continues unabated in cryptorchid animals, so libido, mounting and mating behaviour may appear perfectly normal in cryptorchids.

Among the farm species, cryptorchidism is relatively common in stallions and boars but seen only infrequently in bulls and rams. It is readily detectable, if not by the sight of a small and undistended scrotum, then by careful palpation. This should be undertaken by a veterinary surgeon since normal animals may raise their testes close to the body wall, especially when frightened, and there is then the risk of an incorrect diagnosis. Although cryptorchidism is stated in many texts to be a heritable trait, there are limitations to this statement since the condition is usually bilateral: sterile males will clearly be unable to transmit the defect.

Subfertility in males may result from the effects of heat stress, and of mineral or vitamin deficiencies. High ambient temperatures leading to elevated body temperatures will certainly influence the seminiferous epithelium, in due course reducing the proportion of fertile spermatozoa in the ejaculate. There is usually a delay of 2–3 weeks for this effect to be manifested in bulls, rams and boars – a period equivalent to the transit time of spermatozoa from the seminiferous tubules to the tail of the epididymis. Although bulls, rams and boars are all susceptible to temperature stress, the effects are well illustrated in boars which will produce markedly smaller ejaculates containing reduced sperm concentrations and an increased incidence of morphological abnormalities – especially of the sperm head. Where stud males are housed, it is therefore essential to prevent overheating of the buildings.

Deficiencies of minerals such as copper, zinc and manganese or of vitamins such as A and E also lower male fertility, but these problems should not be common under European conditions of animal production. The effect of such deficiencies is usually temporary in mature animals, but the seminiferous epithelium of young stock may suffer permanent damage from a severe deficiency.

Overworking of males, either in natural service or at artificial insemination centres, may lead to ejaculates containing significantly lower concentrations of spermatozoa and a significantly higher proportion of incompetent immature spermatozoa; these in turn will predispose lower conception rates. Suggested breeding frequencies that should avoid these problems in fully mature males are given in Table 5.3.

Table 5.3 Routine collection frequencies from males standing at artificial insemination centres

Species	Usual number of collections per week
Bull	4–5
Ram*	12–15
Boar	3–4

*Collections are best made only during the breeding season.

Freemartins, intersexes and chromosome anomalies

Further causes of infertility are associated with animals possessing anomalies of the gonads and/or the reproductive tracts. Freemartins have been noted in cattle, sheep and goats, but are most common in cattle and concern infertility in females that have been co-twinned to a male (Fig. 5.4). The female genital tract is rudimentary or poorly developed, and part of the duct system such as the vagina, cervix or uterus is usually blind (i.e. obstructed or imperforate). The precise causes of the freemartin condition are still under debate, but in circumstances where the placental circulations of neighbouring male and female foetuses have fused, androgens and possibly non-steroidal factors from the precociously developing male gonads exert an inhibitory influence on the development of the duct system in the female foetus. The male foetuses develop normally into fertile offspring, but 9 out of 10 females are invariably sterile. If destined for the beef market, the condition of the reproductive tract is clearly of no consequence. If required for replacements in the breeding herd, then it is essential that all female calves born co-twin to a male receive a thorough clinical examination or are excluded. Strange as it may seem, the freemartin condition is exceptional in pigs, probably because fusion of neighbouring placental sacs is extremely rare.

Intersexes are usually genetic females that show a

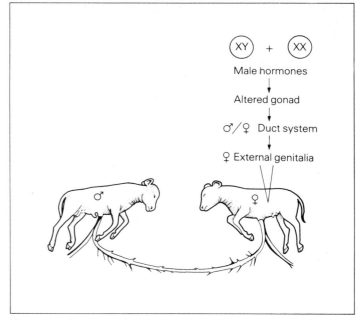

Figure 5.4 Freemartin foetus in cattle. This is a condition in which a male is co-twinned to a female, the placental circulations have fused, and the reproductive system of the female is incompletely developed. Such females are frequently sterile

considerable degree of masculinisation in the reproductive tract, the gonads, and in sexual behaviour. At least one of the gonads appears as an ovo-testis, the adjoining oviduct is rudimentary, and testicular and/or epididymal tissue can frequently be discerned (Fig. 5.5). Such animals are usually especially aggressive after puberty, and will demonstrate many components of male behaviour when placed with oestrous females. Intersexes arise most often in pigs, and their incidence is definitely increased by inbreeding. Intersex animals tend to be sterile, but the condition in pigs may develop only unilaterally, enabling animals to have oestrous cycles, ovulate, and to produce a small litter. Apart from behavioural considerations, intersex females may be recognised by a small vulva, a prominent clitoris, an enlarged preputial sheath, and occasionally some development of a scrotal sac. Incipient tusks are often noticeable in intersex pigs. Culling is again recommended.

Specific anomalies of the chromosome complement are also found in a very small proportion of farm animals that fail to breed, although these can only be revealed by the technique of karyotyping – the preparation of chromosome spreads from individual cells. Instances are well known of bulls, rams and boars with an extra X chromosome, giving a sex chromosome complement of XXY instead of XY. The semen of such males is nearly always devoid of spermatozoa.

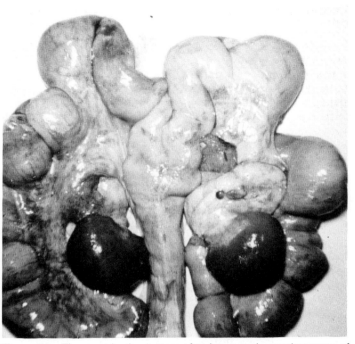

Figure 5.5 The reproductive system of an intersex pig, another cause of infertility or sterility due to abnormal development in the foetus

Oestrogenic plants, alkaloid poisoning

Fertility may be depressed by ingestion of unsuitable feedstuffs, and the period of infertility can extend for weeks or even months beyond the time of consuming the particular feedstuff. A considerable list of possible contaminants could be made, but several examples will be sufficient to illustrate the problem.

Grazing of pastures containing natural oestrogenic substances – which are found especially in certain strains of clover in Australia and in spring grass in Britain – may lead to hormonal disturbance and temporary infertility. Extensive studies in Australia have examined the rôle of subterranean clover in diminishing conception rates in sheep. The major adverse influence is apparently on the processes of sperm transport and storage in the female tract, partly through an alteration in the properties of cervical mucus and also through reduced contractile activity in the uterus.

Other plants can also have anti-fertility effects as shown by ingestion of sufficient quantities of ragwort, broomweed, lupins, onion grass and even pine needles. They may act to prevent conception or by inducing abortion. Grain or straw contaminated with ergot (*Claviceps purpurea*) or a number of other funguses can rapidly produce mycotoxins, terminate a pregnancy, and be followed by a period of infertility.

In situations such as these, infertility will usually be revealed as a herd or a flock problem rather than as instances of breeding difficulties in individual animals. Suspected infertility of this nature is therefore a serious problem, calling for rapid veterinary investigation if major economic setbacks are to be avoided.

Inbreeding and non-specific causes of infertility

The deleterious influence of an excessive degree of inbreeding on fertility, litter size, and even normality of the offspring is widely appreciated. The critical point is to recognise when introduction of new genetic material is necessary. This remains more of an art than a science.

Stress associated with overcrowding of mature females may depress fertility and cause animals in advanced stages of gestation to abort. Although there is no specific evidence so far, odours associated with intact male farm animals may compromise the early (pre-attachment) stages of pregnancy in females mated to a *different* stud bull, boar or ram. There is certainly evidence in laboratory rodents that changing the stud male leads to interruption of early pregnancy due to some form of male pheromonal activity. If a parallel phenomenon can be demonstrated in farm animals, there will be sound reason to pay particular attention to the sequence of use of males in intensively housed units and possibly to avoid changing the stud male when the bulk of the females are already pregnant. This speculation clearly does not apply to situations in which breeding is by means of artificial insemination.

Finally, there is the problem of animals culled from the breeding herd as apparently infertile that subsequently conceive when maintained for examination at veterinary investigation centres or research stations. Fertility in these situations is frequently seen in the absence of any specific treatment, apart from rectal palpation of the ovaries and reproductive tract and sampling of jugular blood. It remains debatable whether veterinary investigation – as distinct from treatment – on the farm would have relieved the problem, whereas the result found upon moving the animals is open to various interpretations. Some local form of stress may have been overcome; there may have been small but critical changes in the diet, or the act of physically moving the animal to a new location may have stimulated ovarian follicular growth and oestrus. There is sound evidence in pigs that transporting and relocating animals will influence ovarian activity, possibly in a manner basically similar to the upset in menstrual cyclicity when primates are moved. Secretion of hormones from the adrenal cortex is thought to be an important factor in this syndrome, and it may well provide an explanation in cattle also.

We are therefore forced to conclude that a proportion of females considered infertile on the farm are simply showing temporary abeyance of successful breeding activity. In the light of the above discussion, the importance of maintaining accurate records for analysing breeding efficiency in individual animals can no longer be denied. Moreover, with the convenience of modern methods of data handling, inspection of such records in conjunction with clinical examination would seem essential before implementing a culling strategy.

Reproductive technology

6

This chapter describes a wide range of techniques for the control of reproductive processes and explains the underlying physiological principles. In no sense, however, does it prescribe methods to be followed rigidly in specific situations. Treatments will be used as considered appropriate to the various husbandry systems in the light of professional advice and the experience of colleagues. Even so, some stockmen must always be in the vanguard of technical advances, and here ultimate success depends not only on motivation but also on adequate understanding, preparation and implementation.

Synchronisation of oestrus

Synchronisation of oestrus or controlled breeding are terms to indicate the process of bringing groups of animals into heat together in response to some form of treatment. Such animals should therefore conceive at closely similar times, proceed through pregnancy together, and produce their offspring in a compact period. The advantages of such a scheme are essentially managerial.

First and foremost, a programme of oestrous synchronisation enables the farmer to regulate the time of heat and breeding rather than permit females to impose their intrinsic reproductive rhythms on the farming system. Some would say that it is manifestly absurd in the 1980s to be checking females once or twice daily for heat if the occurrence of this event can be controlled without affecting conception. In fact, the more recent approaches to oestrous synchronisation not only regulate heat but also enable the moment of ovulation to be predicted, thereby permitting insemination at the optimum time – possibly arranged several weeks in advance. Other advantages of synchronisation include feeding animals in uniform groups with diets appropriate to their stage of pregnancy; supervising birth to reduce neonatal mortality and to arrange cross-fostering; scope for batch weaning, fattening and marketing of animals; and as a general management aid, to rationalise the use of labour, buildings and other resources.

Manipulation of ovarian physiology

The oestrous cycle has been noted to consist of well-defined luteal and follicular phases (Ch. 1, p. 14), and in farm animals the luteal phase is always at least twice as long as its follicular counterpart (Table 3.6, p. 68). In practical terms, synchronising heat in groups of animals involves two possible approaches to manipulating the oestrous cycle, but both rely on the short duration of the follicular phase that follows a specific treatment. The first approach is by inducing regression of the corpus luteum so that all animals in an appropriate group enter the follicular phase and return to oestrus at a closely similar time (Fig. 6.1). The second approach is less direct, and involves suppression of ovarian follicular development during an artificially extended luteal phase so that after removing the hormonal or pharmacological blockade, animals rebound into a compact follicular phase followed by a synchronised oestrus.

Ideally, the form of treatment should be capable of synchronising oestrus in a large group of females in which individual cycles are distributed at random. It therefore follows that in the approach of regressing the corpus luteum, the method must also be capable of dealing with those animals not possessing such a structure at the time of initial treatment; the inference here is that a double treatment may be necessary to ensure that all animals are reprogrammed. On the other hand, the second approach of artificially extending the luteal phase must mean a treatment whose duration is in excess of the luteal phase for that species, if all animals in a random group are to rebound into a follicular phase after treatment.

Cattle

Ever since experiments were started in 1940, the most frequent means of attempting to synchronise oestrus in cattle has involved treatment with some form of progesterone in order to mimic the activity of the corpus luteum. Treatments have consisted of injecting a solution of progesterone, feeding synthetic forms of progesterone (i.e. oral progestagens), implanting silicone rubber capsules of progesterone under the skin, or inserting intra-vaginal sponges or coils containing

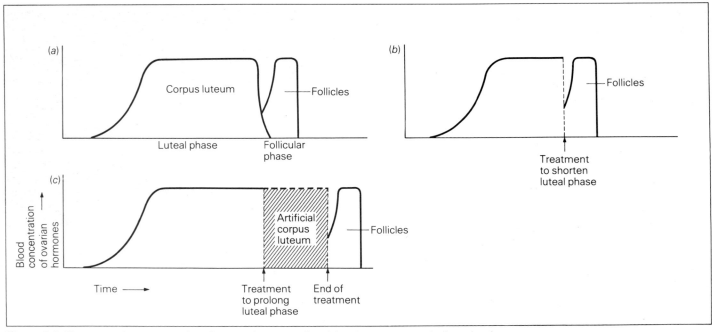

Figure 6.1 Diagrammatic representation of approaches to synchronising the oestrous cycle. Luteal and follicular phases of normal duration are shown in (a), and an abruptly shortened luteal phase is shown in (b) – as would occur after enucleation of the corpus luteum or an appropriate treatment with prostaglandins. Artificial extension of the cycle is shown in (c), as would occur when treating with progestagens

progestagens (Fig. 6.2). The net result has been that animals are placed in an artificial luteal phase with treatment lasting 18–20 days. Synchronisation of oestrus has usually been good with most animals returning to heat 3–5 days after the end of treatment. Fertility, by contrast, after an 18–20 day progesterone or progestagen treatment has been uniformly poor, falling to conception rates of 40–45 per cent or less at the synchronised heat although returning to levels characteristic of the herd (e.g. 60–65%) at the subsequent spontaneous oestrus. A general conclusion from these studies is that any approach involving progesterone or progestagen therapy for a period longer than the luteal phase is likely to depress conception rate at the synchronised heat. On the other hand, progesterone therapy for a shorter period such as 9–12 days does not have this deleterious effect, but a more complex form of treatment is required to control the cycles of a random group of animals.

Many basic studies in the last few years have concentrated on the mechanism whereby the corpus luteum is caused to regress spontaneously at the end of the luteal phase if mating has not been successful, and the key factor has been identified as uterine secretion of prostaglandin $F_{2\alpha}$ ($PGF_{2\alpha}$). Accordingly, treatment with this hormone, or with synthetic analogues of $PGF_{2\alpha}$ that survive better in the circulation, has been used as an alternative approach to synchronising oestrus. Judged from the results of extensive field trials, $PGF_{2\alpha}$ analogues have been remarkably successful in inducing a

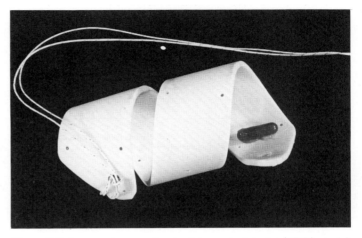

Figure 6.2 A progesterone-releasing intra-vaginal device (PRID) used for modifying the oestrous cycle in cattle. The capsule within the silastic coil contains a solution of oestrogen and progesterone that rapidly melts to give control of the cycle

fertile oestrus when cows have been in good condition and gaining weight. In a group of animals with randomly distributed cycles, the one snag is that a $PGF_{2\alpha}$ injection will only work in about the 55 per cent of cows having a mature corpus luteum. Those with a corpus luteum younger than five days, or those already in the follicular phase, will not respond

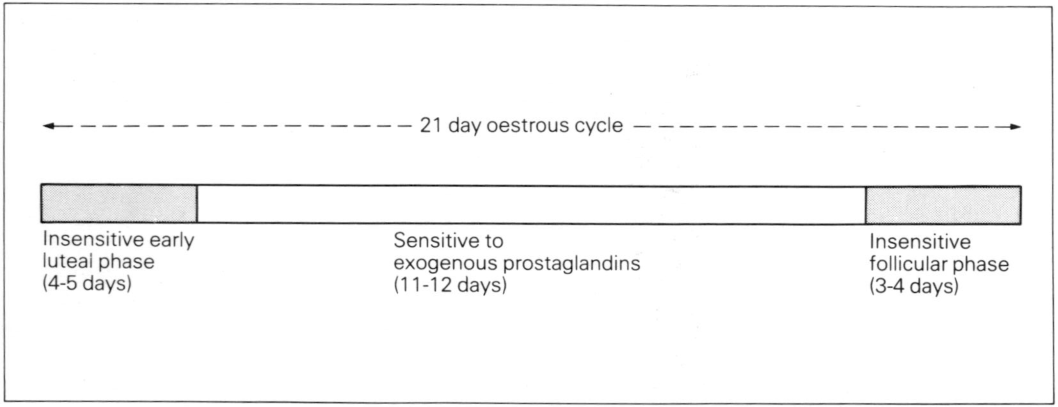

Figure 6.3 The distribution of sensitive and insensitive phases to prostaglandin treatment within the 21-day oestrous cycle of cattle. Because of the insensitive phases, a double-injection regime is required if all animals are to be influenced by prostaglandin treatment

to the injection (Fig. 6.3). However, a second injection given 10–12 days later should find 90–95 per cent of animals with mature corpora lutea sensitive to the lytic influence of $PGF_{2\alpha}$. The animals will move rapidly into a follicular phase and return to heat 2–3 days after the injection, and conception rates at this time should be typical of the herd.

Prostaglandin analogues for synchronising heat in cattle are now marketed by several companies, although available only through the veterinary profession. Because of the prompt and closely synchronised return to heat after the second injection, treatment can be followed by fixed-time insemination without reference to behavioural signs. Since ovulation occurs about 24 hours after onset of the induced oestrus, the recommendation is to inseminate once, 60–72 hours after treatment, or twice, 72 and 96 hours after treatment. Double insemination gives better conception rates, which should approach those found after spontaneous breeding in the herd.

Because of the relatively high cost of two injections of prostaglandins, some effort has gone into developing a short-term treatment based on progestagens that would not depress fertility. Since the duration of treatment (9–12 days) is less than that of the luteal phase, some means is required of regressing a developing corpus luteum whose 'cyclic' lifespan would normally be of approximately 17 days. This is achieved by administering oestrogen at the same time as the progestagen, most conveniently as a solution within a gelatin capsule attached to the inside of a vaginal coil – epithelial temperature melts the capsule and releases the oestrogen solution.

Sheep

Approaches to controlling the time of breeding in sheep follow the same general principles as in cattle. Most of the methods already discussed have been tried, but only two have found extensive application. The first is with intra-vaginal pessaries made of polyurethane sponge material and soaked in a solution of synthetic progestagen. After dusting with antibiotics, these are inserted into the anterior vagina and left there for 14–16 days to simulate the luteal phase. Oestrus will occur spontaneously in most animals 2–4 days after sponge withdrawal. If a single injection of gonadotrophin (e.g. 350–800 i.u. PMSG) is given at the time of withdrawal, animals have a slightly earlier and more compact period of oestrus. Conception rates depend on the time of year and on the number of experienced rams available. Most trials have noted some depression (5–15%) of conception rate at the synchronised oestrus, although this may be alleviated by the PMSG injection and double mating or insemination.

Prostaglandin analogues have also been used to synchronise oestrus, and in sheep the corpus luteum is sensitive to the lytic influence of these hormones from day 4 to day 14 of the cycle. Thus, two injections spaced 8–9 days apart should cause most animals to be in heat two days after the second injection. Fertility at this oestrus appears normal.

Pigs

There has been less demand for methods of synchronising oestrus in this species, since this can be achieved in sows simply by weaning the piglets. Even so, much experimentation has gone into synchronisation procedures in gilts, but a consistent finding is that hormone treatments involving steroids such as progesterone lead to the formation of cystic ovarian follicles and consequent subfertility or infertility. Prostaglandins, on the other hand, have little value in cyclic gilts or sows since the corpora lutea will only regress if treated between day 12 and day 15 of the 21-day oestrous cycle. Other non-steroidal regulators of the cycle (e.g. the drug methallibure) have shown promise, but because they may cause skeletal abnormalities in the foetus – and with the risks to pregnant women rather than pigs in mind – marketing restrictions have been imposed.

As of writing, field trials have just been completed with a new orally active progesterone derivative termed 'allyl trenbolone' (RU-2267). When fed to gilts or sows for 18 days at 20–40 mg per animal per day, oestrus followed 4–6 days after treatment with apparently normal fertility and no development of ovarian cysts.

As a general conclusion, it could be argued that the advantages of synchronisation of oestrus remain considerable, and that animal productivity overall could be improved by a careful application of suitable procedures. Even so, the probable cost/benefit ratio will have to be calculated for specific farming situations; this is likely to be most attractive where controlled breeding would form part of a well-integrated system of production.

Control of ovulation

As is the case for synchronisation of oestrus, control of ovulation focuses on manipulation of the ovaries. But rather than being concerned with the interplay between luteal and follicular lifespans, and thereby with modification of the length of the oestrous cycle, control of ovulation aims to regulate the precise *time of ovulation* and/or the *number of follicles ovulating*; these objectives are achieved most directly by injection of gonadotrophic hormones. The principal reason for controlling ovulation time is that it should enable artificial insemination at the optimum time (with or without detection of oestrus), thus improving conception rates and avoiding the deleterious effects of ageing of the gametes. It should also enable accurate calculation of the time of fertilisation and the developmental stage of embryos which are necessary, for example, in transplantation studies. Modification of the number of follicles ovulating would be valuable when trying to influence fertility directly or by means of embryo transplantation. While control of ovulation time is not difficult to achieve in farm animals, control of the number of follicles ovulating after hormone injection is a notoriously elusive objective: the response of individual animals to a given treatment is unpredictable.

Injection of hormones

Hormone preparations for influencing ovarian follicular development fall into two main categories: (a) those rich in FSH activity that are injected to stimulate the number of follicles developing, and (b) those rich in LH activity used to regulate the time of ovulation. Although the anterior pituitary gland is the classical source of the gonadotrophic hormones (FSH and LH), it is no longer considered commercially convenient, for not only is collection of large numbers of pituitary glands at the slaughterhouse involved, but also extraction and preparation of material in a condition suitable for storage and subsequent injection. The placentae of various species also develop the ability to secrete hormones having FSH- or LH-like activity, and preparations of placental gonadotrophins have found widespread use in treatments for regulating ovulation in farm and laboratory animals. Two preparations predominate: (a) pregnant mare's serum gonadotrophin (PMSG), which is rich in FSH activity and is derived from the endometrial cups of the mare placenta (serum concentrations of this hormone are greatest between days 60 and 90 of gestation (Fig. 6.4)); and (b) human chorionic gonadotrophin (HCG), which is rich in LH activity and is derived from the chorionic layer of the human embryo. In contrast to the secretion of PMSG into blood, HCG readily accumulates in the urine (Fig. 6.5) and is prepared commercially from samples responding positively in a pregnancy diagnosis test. The gonadotrophins are concentrated from the respective fluids, freeze-dried, and marketed in glass vials containing a known quantity of FSH or LH. This measurement has been derived from a biological assay of the gonadotrophic potency (based on the ovarian response of immature mice), and is expressed in international units (i.u.).

Specific treatments

The conventional time at which to inject PMSG to increase the *number of follicles maturing* has been at the onset of the follicular phase of the oestrous cycle (Fig. 6.6); that is, immediately upon regression of the corpus luteum when the animal's pituitary gonadotrophins would themselves be stimulating follicular growth (Table 6.1). The injected (i.e.

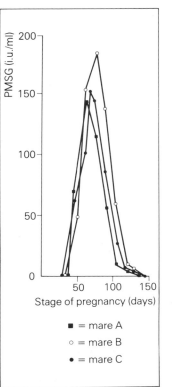

Figure 6.4 The time course of secretion of the chorionic gonadotrophin, pregnant mare's serum gonadotrophin (PMSG), in horses. This usually reaches a peak between days 60-90 of gestation, and is rich in FSH-like activity

■ = mare A
○ = mare B
● = mare C

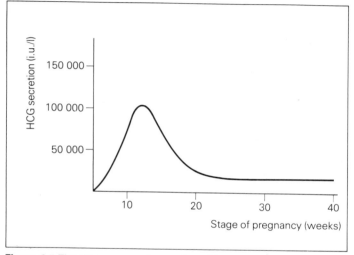

Figure 6.5 The time course of secretion of human chorionic gonado-trophin (HCG) as monitored in its urinary excretion. HCG is rich in LH-like activity

exogenous) gonadotrophin therefore reinforces the influence of endogenous gonadotrophins, permitting a greater number of follicles to achieve the pre-ovulatory diameter. The action of PMSG should not only increase the number of follicles ripening but may also slightly accelerate their terminal

Table 6.1 The onset of the follicular phase relative to the duration of the oestrous cycle, and the conventional time to inject PMSG in order to increase the number of follicles ovulating

Species	Day on which follicular phase begins	Mean duration of oestrous cycle (days)
Cow	17/18	21
Sheep	14/15	16.5
Pig	15	21

growth. In this situation, heat occurs a day or so sooner than anticipated due to increased secretion of follicular oestrogens.

Examples of quantities of PMSG considered appropriate to the farm species are listed in Table 6.2; the dose selected is given as a single subcutaneous – or less often intramuscular – injection. This placental gonadotrophin will continue to

Table 6.2 Examples of doses of PMSG and HCG (in international units) used for regulating the extent and timing of ovulation

Species	Dose of PMSG* (i.u.)	Dose of HCG† (i.u.)
Cow	2000–3000	500–2500
Sheep	500–800	250–500
Pig	750–1500	500–1000

*PMSG is injected subcutaneously or intramuscularly.
†HCG is injected intramuscularly or intravenously.

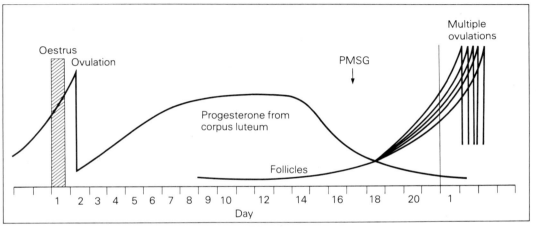

Figure 6.6 The optimum time at which to inject PMSG to seek a superovulatory response is at the beginning of the follicular phase when the secretion of progesterone would have waned and Graafian follicles should be actively maturing

stimulate ovarian follicles over a period of several days, and no advantage is obtained by splitting the dose into a series of injections. If the response to the injection in terms of the number of follicles ovulating is significantly above the normal ovulation rate for the species or breed in question, then this is referred to as *superovulation*. Apart from experimental studies on eggs and fertilisation, such as *in vitro* fertilisation, the principal use of superovulation in farm animals is to provide embryos for transplant studies (see below).

Another method of using PMSG has been to inject animals during the mid-luteal phase of the oestrous cycle, and then to regress the corpus luteum with PGF$_{2\alpha}$ or an analogue one or two days after gonadotrophin treatment. The rationale here is that there may be a more uniform although smaller sized crop of follicles available for stimulation earlier in the oestrous cycle, and treatment may therefore produce a more consistent response. Although convincing evidence for the latter is still awaited, this dual treatment for securing superovulation is favoured by many cattle embryo transplant units. It most commonly takes the form of 2000 i.u. PMSG on day 11 or 12, followed by 750 μg cloprostenol (Estrumate, ICI) 48 hours later. Ovulation itself will occur without further hormonal treatment, although the period required for all the stimulated follicles to ovulate may be protracted. This, in turn, may lead to poor fertilisation, although double insemination with increased numbers of spermatozoa is beneficial.

A single injection of HCG, administered intravenously or more commonly intramuscularly, is used for controlling the *time of ovulation* – that is, the moment of follicular collapse with release of the egg(s). The timing of the response after injection parallels the time of ovulation after the onset of heat (Table 6.3), although the accuracy of the estimate would depend on the accuracy of detection of the onset of heat. But it is known that there is a peak release of LH and FSH very close to the onset of heat in farm animals, and that the interval from this peak of gonadotrophic hormone release to ovulation is remarkably constant (Fig. 6.7). The objective in controlling ovulation with HCG is to inject the animal some hours, *not* days, before the endogenous release of gonadotrophins, so that the mature follicle(s) is programmed by and responds to the injection. In reality, there is a fine judgement in waiting

Table 6.3 Timing of ovulation in response to an injection of HCG given during pro-oestrus and thus before the surge of endogenous gonadotrophin secretion

Species	Interval from HCG injection to ovulation (hours)
Cow	30
Sheep	26
Pig	40

late enough in pro-oestrus to have a fully mature follicle before injecting HCG, and waiting too long such that the animal may already be in oestrus and therefore have released her own surge of pituitary gonadotrophins. In the latter instance, control of ovulation time is not obtained. One way around this dilemma is to inject HCG at a known interval after PMSG stimulation or after an oestrous synchronisation treatment – for example, after removing intra-vaginal sponges. Doses of HCG injected vary between 250 and 2500 i.u. (see Table 6.2).

Embryo transplantation and storage

The technique of egg transplantation, or more correctly embryo transplantation, has developed during the last twelve years from a research procedure to one with a significant

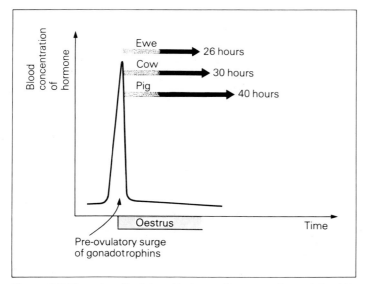

Figure 6.7 To portray the interval between the surge of gonadotrophin secretion found at the onset of oestrus and ovulation of the follicle(s) 26-40 hours later according to species

commercial impact. Its application has been largely in cattle – some 17 000 pregnancies were produced by transplant procedures in North America in 1979 – but sheep, pigs and horses have all received attention. The technique requires recovery of embryos by flushing fluid through the reproductive tract of the donor animal (which may or may not have been superovulated), examination of the embryos under a binocular microscope, and then their insertion into the reproductive tract of the recipient or foster mother using a glass pipette (Fig. 6.8). The recovery and transplantation procedures may involve abdominal surgery under full or local anaesthesia, or they may gain access to the uterus of conscious animals through the vagina and cervix – as in artificial insemination. Although the necessary procedures have now reached a high technical standard, and there is the added possibility of performing a range of manipulations on the embryos between recovery and transfer, there is nothing novel in embryo transfer itself. Viable offspring were obtained in this manner from a rabbit in the late nineteenth century.

Three important biological facts underlie the success of embryo transfer in farm animals. First, attachment of the developing embryo(s) to the wall of the uterus does not occur for some two weeks in pigs and later than three weeks in cattle (see Ch. 3, p. 61). During this interval, the embryo is sustained largely by secretions of the uterus, and because it is free-living in a fluid environment it can be removed and transplanted in a culture medium without harm. Second, within a species the transplanted embryo is not rejected by the reproductive tract of the recipient, despite the fact that it represents a foreign tissue. Perhaps this is not so surprising as

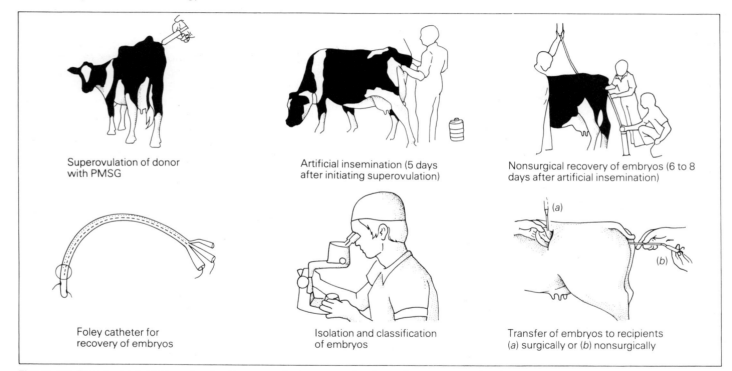

Superovulation of donor
with PMSG

Artificial insemination (5 days
after initiating superovulation)

Nonsurgical recovery of embryos (6 to 8
days after artificial insemination)

Foley catheter for
recovery of embryos

Isolation and classification
of embryos

Transfer of embryos to recipients
(a) surgically or (b) nonsurgically

Figure 6.8 The sequence of stages involved in non-surgical recovery of embryos from a superovulated
donor cow and their transfer to a recipient by surgical or non-surgical means

at first sight, for 50 per cent of the genetic complement of any embryo – the paternal contribution – must be foreign to the mother. The third point is that development of an embryo within the reproductive tract of a recipient has no detectable effect on its genetic constitution, and this is clearly of fundamental importance to pedigree breeders. Thus, Simmental embryos transplanted to Friesian recipients yield pure Simmental calves. On the other hand, the recipient will influence the birthweight of the calf, and will also confer on it valuable immunity against disease.

The following sections concentrate on some essential points in the application of transfer techniques to cattle, since this is likely to remain their major use in the foreseeable future. The underlying principles apply to the other farm species, although points of detail must necessarily differ.

Supply of embryos

A supply of suitable embryos is essential, and procedures of superovulation are usually applied to the donor animal. These may take the more conventional form of PMSG injection on approximately day 17 of the oestrous cycle or one given during the luteal phase (e.g. days 9–12), 48 hours after which the corpus luteum is caused to regress and oestrus induced by treatment with a prostaglandin analogue. The superovulated donor is bred naturally or, if by artificial insemination, with an increased number of spermatozoa and usually two or three inseminations 48 and 72 hours after prostaglandin injection. Where the transplantation technique is being used to propagate a genetically superior donor (e.g. a valuable pedigree animal), and thereby to increase her rate of reproduction (Fig. 6.9), it is clear that the supply of embryos must come from this individual. Alternatively, if transplantation is being used simply to increase the supply of beef calves, perhaps by induced twinning, then mature non-pedigree stock can be treated. Thus, heifers of beef type such as Hereford–Friesian crosses may be superovulated and inseminated a few days before slaughter, the reproductive tracts recovered in a sterile flask at the abattoir, and the embryos flushed from the tract in the laboratory. In the case of the pedigree donor, on the other hand, recovery of embryos must involve surgical or non-surgical flushing of the tract.

Embryos may be transplanted in the fresh condition shortly after examination under a dissecting microscope, or they may be used after storage on a temporary or long-term basis. The advantage of storage is that it confers flexibility; this is in terms of avoiding wastage of embryos if there is a particularly good superovulatory response and the number of embryos exceeds the number of suitable recipients. Availability of stored embryos also permits greater flexibility in the timetable of events. Cow embryos are now beginning to be stored routinely in liquid nitrogen at $-196\,°C$.

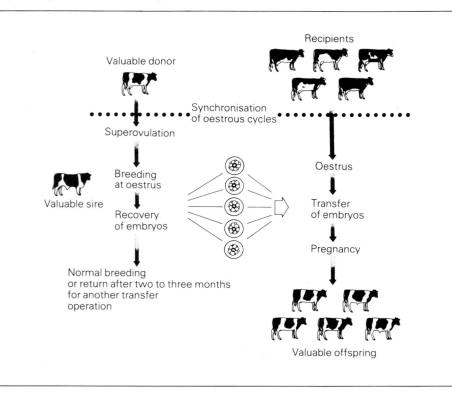

Valuable donor

Recipients

Synchronisation
of oestrous cycles

Superovulation

Valuable sire

Breeding
at oestrus

Oestrus

Recovery
of embryos

Transfer
of embryos

Pregnancy

Normal breeding
or return after two to three months
for another transfer
operation

Valuable offspring

Figure 6.9 Advantages of embryo transplantation illustrated schematically. Embryos from a valuable donor mated to a valuable sire are transplanted to non-pedigree recipients that act as foster mothers for development of the pedigree calves

Synchronisation of donor and recipient

Successful transplantation of embryos depends on close synchronisation of the oestrous cycles of donor and recipient animals, because (a) growth of the embryo is tightly programmed in terms of its requirements from the uterine secretions, and (b) the chemical nature of these secretions is changing progressively with the time elapsing from ovulation. Thus, an embryo transplanted to a non-synchronous reproductive tract will not be bathed by secretions necessary for its stage of development. In practice, there is some latitude in the requirement for synchronisation and many studies have suggested that ±1 day in cattle and ±2 days in sheep are still compatible with establishment of pregnancy (Fig. 6.10). Nonetheless, the best conception rates are obtained when the oestrous cycles of donor and recipient are precisely synchronised (Table 6.4). This can be achieved by having a sufficiently large pool of recipients so that, by chance alone,

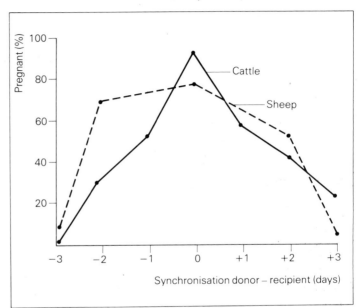

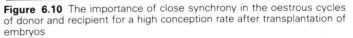

Figure 6.10 The importance of close synchrony in the oestrous cycles of donor and recipient for a high conception rate after transplantation of embryos

Table 6.4 To illustrate the importance of oestrous cycle synchronisation to the success of embryo transplantation procedures in cattle

Variation from exact synchronisation (days)	No. of recipient animals	Diagnosed as pregnant (%)
0	25	90
±1	50	55
±2	20	35

some animals should be in heat at the same time as the donor animal. Alternatively, the oestrous cycles of prospective donor and recipients can be synchronised by treatment, such as by injection of a prostaglandin analogue. This approach certainly aids forward planning, and reduces the very considerable number of recipients that would need to be maintained in the former scheme – stated to be >200 animals for a daily programme of transplants. As inferred above, the use of stored embryos is a third means of meeting the requirement of synchronisation, since embryos can be taken from storage when recipients have reached a stage of the cycle appropriate for transfer.

Site of recovery and transfer

Although early cleavage stage embryos can be recovered quite successfully from the oviducts during abdominal surgery, it is more convenient in most circumstances to wait until the embryos have entered the uterus, three days or so after ovulation. Once the embryos have passed through the utero-tubal junction, the stage of flushing has been dictated largely by the conception rate achieved following transplantation: embryos aged 6–8 days are currently considered ideal (Fig. 6.11). The excellent conception rates obtained with such embryos must in part be a reflection of the morphological selection of embryos before transfer, those appearing as healthy developing blastocysts being transferred while eggs or embryos of retarded or fragmented appearance are discarded.

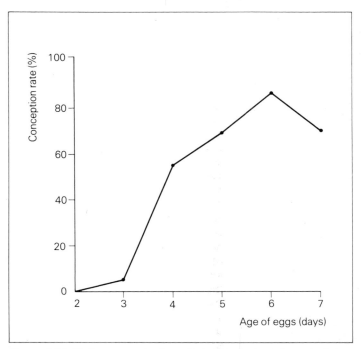

Figure 6.11 Optimum conception rates after embryo transplantation procedures in cattle are obtained with 6-day old blastocysts. Selection of embryos to be transplanted undoubtedly contributes to the success rate at this stage

Recovery can be by surgical or non-surgical procedures. In either case, the uterine lumen is flushed with a sterile physiological solution that is collected through a rubber catheter with a balloon cuff into a round-bottomed glass dish (Fig. 6.12). Whereas a surgical flush through a flank or mid-ventral abdominal incision would use approximately 50 ml of medium per uterine horn, a non-surgical flush would use 200–300 ml introduced via the cervix with the uterus being massaged *per rectum*. The medium is at body temperature and must subsequently be protected against temperature shock and bacteria during examination under the microscope. The embryos should collect in the bottom of the glass dish, and those to be transferred are placed in fresh medium before aspirating into the transfer pipette.

In the case of a surgical transfer, a fine Pasteur pipette is used and enters the lumen of the uterus through a puncture wound made with a needle; the embryo contained in a droplet of medium is then expelled. This can be done through a flank incision under a local anaesthetic with the sedated cow standing in a crush. Conception rates obtained by this procedure are of the order of 55–65 per cent. Non-surgical transfer involves passage through the cervical canal, and this in turn requires a catheter of suitable dimensions. Procedures are not yet as reliable as those of surgical transfer, this being reflected in conception rates of 40–55 per cent. Non-surgical recovery of embryos from suitable donor cows is theoretically possible every 40–50 days, perhaps six times per year. The yield of embryos is extremely variable (e.g. 0–35 embryos), but an average yield is 4–6 embryos per recovery.

If a single embryo is being transplanted to an unbred recipient, it must be introduced into the horn of the uterus adjoining the ovary with a corpus luteum. If an embryo is being added to an animal already mated or inseminated, then this is deposited in the uterine horn opposite the ovary with the corpus luteum. In the case of transplantation of two embryos, it is essential to deposit one per horn since there is little intra-uterine migration in cattle, and competition between growing embryos may lead to death of one or both. Experimental observations suggest that two embryos in one uterine horn will yield <20 per cent twins, whereas one embryo per horn can give >70 per cent twins.

Uses of embryo transplantation

Perhaps the two most attractive uses of embryo transplants are (a) rapidly multiplying up the number of offspring from a pedigree or exotic donor (Fig. 6.9, p. 118) and (b) increasing total calf output by means of twinning – the heritability of twinning being less than 10 per cent. Other uses of the technique are suggested in Table 6.5. The potential profit involved in pedigree breeding may justify use of a surgical approach to embryo transplantation, not least since the proportion of embryos recovered and the conception rate subsequently obtained should both be higher than after

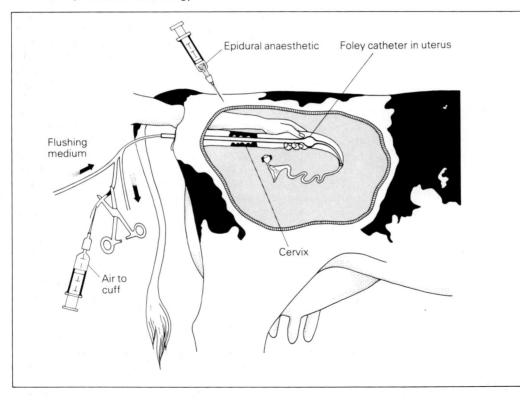

Epidural anaesthetic

Foley catheter in uterus

Flushing medium

Air to cuff

Cervix

Figure 6.12 Use of a Foley rubber catheter with inflatable cuff for recovery of embryos from the uterus in cattle. A physiological medium is introduced through the catheter, flushed around the uterine horn, and returns through the catheter to a collection dish

Table 6.5 Some possible uses of embryo transplantation procedures in cattle

1. Produce more calves from best cows in short period;
   i.e. more female calves of high value and
      better bull calves for AI centres
2. Rapid multiplication of rare or commercially desirable breeds
3. In conjunction with superovulation, to speed up selection programmes
4. To induce twinning;
   i.e. either by transplanting two embryos to unmated recipient or by
      adding one embryo to a mated recipient
5. As part of an export programme to upgrade stock in developing countries
6. To shorten the generation interval by breeding from prepuberal females
7. As an experimental procedure, after deep-freezing, cloning or sexing of embryos

non-surgical interventions. Even so, a cow is unlikely to tolerate more than three or four abdominal operations, especially mid-ventral ones, before the problems of weakening of the body wall and adhesions of the reproductive tract become serious; thereafter, she is returned to the herd for 'normal' breeding. The procedures for induced twinning have already been referred to, these being transfer of a single embryo to each uterine horn of an unbred cow or 'topping up' the empty horn of an animal already mated. Both approaches are readily attained by non-surgical methods which, of course, are the only methods likely to have an extensive field-scale application.

Although many beef units suckle several calves per cow, the thought of the mother actually producing twins is not always

welcome and various snags are raised at the suggestion of induced twinning: these can be responded to rationally. The first is that twin calves are of lower individual birthweight than singletons. Clearly this is true, although it can be alleviated to some degree by correct feeding of the dam in the latter half of pregnancy and there should be no stunting effect on the calf's subsequent growth rate. The advantage of twinning is that the overheads of a 9-month gestation are carried by two offspring rather than one. Second, the incidence of retained placenta is higher, and this may require veterinary intervention (i.e. cleansing). The problem is almost certainly due to a slight asynchrony of calf maturity with the more advanced foetus precipitating delivery of its twin, even though the placental cotyledons of the latter may not have fully separated from the wall of the uterus. Once again, appropriate nutrition in late pregnancy should help to reduce the incidence of this condition by promoting a more synchronous growth of the two foetuses. Third is the problem of freemartins when a female is co-twinned to a male and the placental circulations have fused. The reproductive tract of the female shows incomplete or grossly abnormal development, and she will usually be sterile due to the influence of androgens from the precociously developing male foetus. Here the point is, surely, that if induced twinning is being used to increase production of beef calves and the animals are destined for slaughter rather than breeding, then the condition of the reproductive tract is of no consequence.

Accordingly, induced twinning by a non-surgical method of

embryo transplantation should feature more widely in cattle production systems of the 1980s. Embryo transplantation in the sheep and pig industries is likely to remain largely for import and export schemes in order to introduce new breeds or to modify breed types.

Diagnosis of pregnancy

A prime objective of pregnancy diagnosis is to identify those animals that have not conceived to mating or insemination and *to predict their return to oestrus*. In this way, it should be possible to supervise them closely and to rebreed them at the optimum time with the loss of only one barren cycle. Loss of further cycles, of approximately three weeks duration in cattle and pigs, could lead to severe financial penalties under intensive production systems. While a powerful case can therefore be made for methods of pregnancy diagnosis within the length of one oestrous cycle, there is a related snag: the earlier a positive diagnosis of pregnancy is made, the lower its correlation with production of a viable foetus due to embryonic death which may reach proportions of 30 per cent or more. As a sequel to an early pregnancy test, some further procedure may thus seem worthwhile before feeding diets or subjecting animals to management systems appropriate to the later stages of gestation.

Not to be overlooked when considering an early test is that the laboratory procedures must not take so long that they compromise the objective of identifying animals at least a day before expected oestrus. Other requirements of a pregnancy test are that it should be accurate, inexpensive, easily and quickly performed, and involve a once-only handling of the animals (Table 6.6). The extent to which these criteria are met can be judged from the descriptions that follow.

Table 6.6 Some desirable characteristics in methods of pregnancy diagnosis for farm animals

A	Accurate
E	Early
I	Inexpensive
O	Once only
U	Uncomplicated

The most widespread test for establishment of pregnancy in farm animals is the presumptive one in which the mated female fails to return to oestrus, especially as monitored in the presence of a mature male. In such circumstances, which commonly apply in sheep and beef cattle, this test is valuable although it does require a suitable system of marking those females returning since mounting and riding activity may not be seen. Stud or teaser males are raddled or wear a crayon harness in the centre of the brisket. Where artificial insemination is used and a male is not maintained on the unit, detection of a return to heat is a much more chancy affair, and a more positive approach is needed.

In these circumstances, the presence of a gravid uterus can be judged (1) directly from physical detection of the embryo or foetus, (2) less directly from the status of the ovaries and/or their secretion of progesterone as monitored in blood or milk, (3) indirectly from an influence of the embryo and ovaries on the characteristics of the neighbouring blood vessels or vaginal epithelium, or (4) possibly from chemical or immunological detection of embryonic or placental secretions. Of these approaches, that used most widely in cattle, sheep and pigs has usually depended on physical detection of a foetus. Therefore, by definition, this is after the phase of the developing embryo and after its attachment to the uterine wall. If, on the other hand, secretions from the early embryo were detectable in the mother's blood, milk, saliva or urine, these might form the basis of a good predictive test. This, incidentally, is the means of early pregnancy testing in women when HCG from the blastocyst is estimated in urine.

Palpation

The foetus and its placental membranes may be detected by palpation with the hand inserted *per rectum* (Fig. 6.13). This is the traditional veterinary approach in cattle and horses, but it clearly cannot be applied in sheep and pigs due to limitations of size. In cattle, it can be used successfully in the sixth week of gestation, but greater reliability is obtained after day 40. When the foetus cannot be readily detected, it is common practice to manipulate the uterine wall between finger and thumb to feel a slipping or sliding of placental membranes. This must be done with care to avoid precipitating an abortion. The ovaries are also palpated during these procedures to confirm the presence of a mature corpus luteum. Rectal palpation in mares is performed 20–25 days after breeding, with hobbling as a necessary precaution.

There is scope for a different form of palpation for detecting pregnancy in sheep and pigs. Abdominal palpation (ballottement) is possible in ewes after days 60–70 of gestation, usually with the animal on her rump. Alternatively, as sometimes practised in North America, the ewe is examined on its back with a plastic probe inserted in the rectum to displace the gravid uterus towards the body wall. Palpation through the rectal wall for enlarged uterine arteries is possible in sows after day 45 of gestation, whereas changes in the pulsation response of these arteries to digital pressure – resulting in vibration or fremitus – will reveal pregnancy from day 30 onwards. Palpation is eased if food and water are withheld for 12 hours before the examination, but experience and care are still essential.

Ultrasonic scanning

Another method, which gives a direct result and avoids secondary handling of animals, is detection of a gravid uterus by means of a probe that transmits and receives a narrow

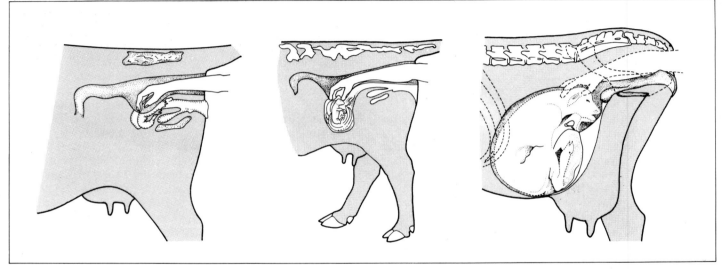

Figure 6.13 Pregnancy diagnosis in cattle by means of rectal palpation for an enlarged uterus and its placental and foetal contents

ultrasonic beam. Once again, however, it is not applicable in the earliest stages of pregnancy nor is it used in cattle. In sheep, it is accurate from the eighth week of gestation; in pigs as early as the fifth. The principle involved is that of reflection of an ultrasonic signal from the moving object back to source at an altered frequency – the Doppler phenomenon. The probe is applied to the abdomen and sound is reflected from foetal heart movements, blood flow in the umbilical cord, or from the fluid in the foetal sacs; it is analysed audibly or converted into a visual display on a small screen (Fig. 6.14). If

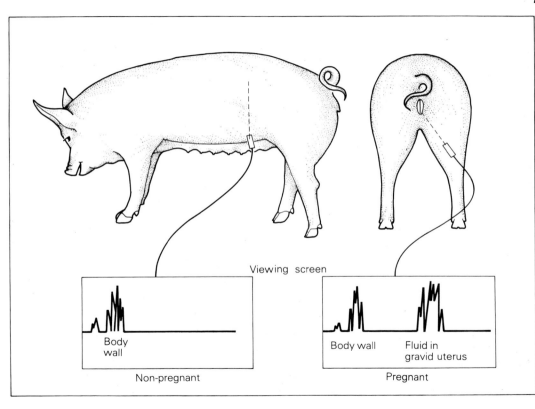

Viewing screen

Body wall

Non-pregnant

Body wall

Fluid in gravid uterus

Pregnant

used in conjunction with a rectal probe, the apparatus can be used in sheep from about day 40 onwards as the umbilical vessels produce a characteristic swishing sound. The main disadvantage of ultrasonic devices is the cost of the equipment.

Radiography

X-irradiation of the uterine contents is successful in sheep after the 50th day when foetal ossification is detectable, and it is very accurate by three months of gestation. Although the equipment is again expensive, there are portable versions and the method enables rapid identification of ewes carrying twins or triplets. Even so, many physiologists have reservations about this approach because of radiation risks to the foetal gonads.

Vaginal biopsy

The removal of a very small sample of vaginal epithelium with a simple scraping device is the basis of a pregnancy test in pigs, applicable between days 20 and 30 after breeding. The number of epithelial cell layers is conspicuously reduced in sows that have conceived (Fig. 6.15), although microscopic examination of the fixed and stained tissue is certainly necessary. Other disadvantages are that some skill is required in performing the biopsy, care must be taken in labelling

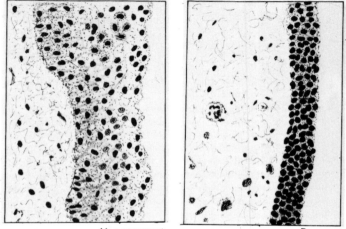

Non pregnant   Pregnant

Figure 6.15 Pregnancy diagnosis in pigs by means of a vaginal biopsy and examination of the state of the epithelium. This is multi-layered in cyclic animals compared with the two or three cell layers of the vaginal epithelium in pregnancy

samples, these must be obtained at a known stage after breeding and there is not an immediate result.

Progesterone measurements

Corpora lutea must be maintained in an active state to support a pregnancy, and measurement of the hormone progesterone secreted by these structures should enable an early diagnosis. While predictable concentrations of progesterone will be circulating in pregnant animals, blood sampling followed by the preparation of plasma or serum is unlikely to be convenient on the farm. On the other hand, the concentration of progesterone in milk parallels that in blood, so there is the possibility, in dairy cows at least, of a pregnancy diagnosis based on milk progesterone concentrations. The Milk Marketing Board in the UK offers a semi-automated service for performing the measurement. In brief, a 5 ml sample of whole milk from the recording jar or milk bucket is collected into a glass bottle containing a preservative, and the labelled bottle is stored at 5 °C until despatch to the central laboratory. Sampling is recommended 22–26 days after breeding, and although analysis is performed within 24–48 hours of receipt and achieves 85–90 per cent accuracy, it still does not normally enable non-pregnant animals to be re-inseminated until two barren cycles (i.e. six weeks) have elapsed. To correct for the influence of embryonic death, confirmation of pregnancy is possible at 15 weeks (105 days after breeding, or later) by means of another hormone test: the concentration of oestrone sulphate derived from the foetal membranes. This test is also provided commercially by the Milk Marketing Board.

The estimation of blood concentrations of progesterone is possible in heifers, but blood sampling has little attraction for pregnancy diagnosis in the other farm species.

Laparoscopy

Visual inspection of the abdominal contents upon inserting an endoscope through a small incision under local or general anaesthesia is referred to as laparoscopy. It can be used to examine the ovaries for an established corpus luteum and the uterus for a swelling characteristic of a developing embryo. In skilled hands, laparoscopy can be applied in the third week after breeding; it is rapid, accurate, and gives a direct result. Even so, it is likely to remain a research tool in sheep and cattle and not a field-scale approach to pregnancy diagnosis; it is rarely used in pigs.

Embryonic hormones

Secretion of hormones by the early embryo might be exploited in pregnancy diagnosis if such hormones could be detected in maternal body fluids. This is certainly the case in women, where chorionic gonadotrophin in the urine forms the basis of a test as early as eight days after fertilisation. Likewise, gonadotrophin in the blood of pregnant mares (i.e. PMSG) is used for diagnosis by day 40 after breeding. There is evidence that the pre-attachment embryos of sheep, pigs and cattle secrete gonadotrophin-like hormones, but a reliable test for revealing these in maternal fluids has yet to be reported. When this step is achieved, pregnancy diagnosis will have reached the ideal situation of being early enough to reveal which animals will return to oestrus and need to be rebred. In the case of a positive diagnosis, it may even become possible to estimate the number of viable embryos by this approach.

As a concluding remark, it is worth noting that diagnosis of pregnancy in itself does nothing to improve herd or flock fertility. In conjunction with a good system of records, though, it is a valuable means of monitoring herd performance and may provide the first warning sign of a fertility problem associated, for example, with infectious disease or ingestion of contaminated feedstuffs.

Induction of parturition

Control of the time of calving, lambing or farrowing would be beneficial in most systems of farming, not least when we consider that problems during or shortly after birth constitute a major source of perinatal loss. An understanding of the mechanisms involved in spontaneous parturition has suggested ways of controlling birth, but these can only be applied fruitfully if the day of mating or insemination is known. Thus, induction of parturition is likely to be restricted to more intensive enterprises where good records are kept. If synchronisation of oestrus has been used prior to breeding, then controlled breeding may be seen as a logical sequel. Successful procedures should eliminate weekend or even night-time deliveries, and facilitate increased supervision of animals in labour, and then of the newborn. Close synchrony in the time of birth would also enable cross-fostering, and overall may be viewed as a management aid. But having suggested some potential advantages of induced parturition, it is worth recalling that the foetus is growing rapidly in the lattermost stages of gestation, and in particular is building up its energy reserves of liver and muscle glycogen in preparation for the rigours of birth. Thus, premature induction could have disastrous consequences and may lead to the related problem of a retained placenta since separation of the foetal and maternal membranes would not have proceeded sufficiently far.

Although parturition can be induced in all the large farm species – and indeed may be necessary as an emergency clinical procedure for overdue birth, toxaemia or a bone fracture in the dam – commercial application is common only in cattle and pigs. Arguments for inducing birth in cattle focus mainly on the seasonal constraints to milk and grass production in countries such as New Zealand, where it is important to have cows calving close to an optimum date – even at the price of significant calf mortality. But herds with a high incidence of calving difficulties (dystocia) may also benefit from induction procedures, as would dairy cows in calf to large beef breeds such as the Charolais. Treatment involves injecting the mother with a corticosteroid drug, the objective of which is to simulate the foetal switch for the termination of pregnancy (see Ch. 4, p. 75). These drugs may be administered either in short-acting form, 2–3 days before the anticipated date of calving, or in long-acting form, 2–3 weeks before calving. The difference here is in the form of preparation: long-acting corticosteroids are suspensions or insoluble esters of the synthetic steroid whereas short-acting versions are in the free alcohol or soluble ester.

Gestation lengths vary from 275 to 290 days in cattle, being a week or so shorter in animals carrying twins. When a synthetic corticosteroid, such as dexamethasone, is injected in the short-acting form after days 265–270 of gestation, calving usually follows within 2–4 days. The golden rule is that the nearer to the anticipated time of calving the drug is given (e.g. 20 mg dexamethasone as a single intramuscular injection), the more likely it is to be successful in inducing a viable birth. Nonetheless, complications are not uncommon and, of these, retained placentae and stillborn calves are foremost. There may also be a depressing effect on milk yield and a delayed return to fertile oestrus. Summarising observations from a number of studies using short-acting drugs, 70–80 per cent of treated animals should calve within four days of treatment, but the incidence of stillborn calves may be 10–15 per cent and that of retained placenta 25–30 per cent.

Use of long-acting (depot) corticosteroids leads to a lower incidence of retained placenta (8–20%) since more time is available for separation of the foetal and maternal components, but stillborn calves are more common (10–25%) and many of these are conspicuously underweight. Moreover, use of long-acting drugs after day 260 may not significantly advance the calving date, whereas use earlier than about day 250 will almost certainly increase the incidence of stillbirths or neonatal mortality. This latter situation may be acceptable to New Zealand farmers, but it is unlikely to be tolerated elsewhere. Long-acting corticosteroids seem not to depress milk yield, but pre-partum milking may become necessary following treatment. In conclusion, therefore, use of either form of drug to induce calving may be justified as a management aid in specific circumstances, but routine use of corticosteroids may compromise both mother and calf. Other forms of hormonal intervention, such as the use of corticosteroids in conjunction with a prostaglandin analogue, or indeed prostaglandins alone, have not been shown so far to mitigate the snags already described, although the moment of calving may be made more precise.

Induction of farrowing has been more of a success story. Because continued secretion of progesterone from the maternal ovaries is essential for the maintenance of pregnancy in pigs, and since the corpora lutea can be caused to regress promptly following a single injection of a prostaglandin analogue, a simple means of precipitating birth is available. Gestation length approximates 115 days in gilts and sows (Fig.

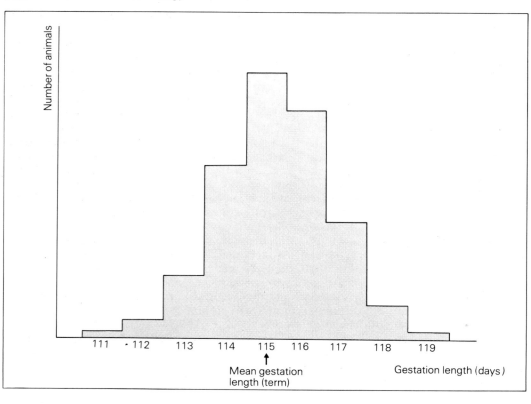

Figure 6.16 The distribution of gestation lengths in pigs with a mean close to 114-115 days

6.16), and if the time of breeding has been recorded, it is a straightforward procedure to advance the time of farrowing by a day or two. Too premature a treatment leads to a steep reduction in the number of viable piglets. The recommendation is to calculate the mean gestation length for the herd, and then to give a single intramuscular injection of a prostaglandin analogue (e.g. 175 μg of cloprostenol analogue; Planate, ICI) two days before the calculated day of farrowing. The majority of animals can be expected to deliver 24 ± 6 hours after the injection (Fig. 6.17), and 90–95 per cent should have commenced within 36 hours of treatment; those animals that have already entered labour spontaneously constitute the exceptions. Because of the precision of the response, induction of farrowing during the working day is a realistic objective attained by injecting the drug between 1000 and 1100 hours. The biggest attraction will be in intensive units, for close planning of birth schedules will aid efficient use of skilled labour and specialist buildings; cross-fostering of piglets should also be possible. In the light of these advantages, and since there are no specific snags when the recommendations are closely followed, induced parturition in pigs would seem set to become a routine husbandry practice, even though prostaglandin analogues are only available commercially via the veterinary profession.

Treatments also exist for regulating the time of lambing, with the customary proviso that they only work satisfactorily when applied at a known stage of gestation. As is the case in cattle, synthetic corticosteroids rather than prostaglandins are favoured. Gestation lengths in sheep usually fall between 144 and 153 days, and 16 mg of short-acting dexamethasone given as a single injection on day 144 will induce lambing approximately 45–48 hours after treatment; 75 per cent of the lambing should be concentrated within a period of 24 hours. At a lower dose, such as 12 mg dexamethasone, the response is delayed until about 52–55 hours, and this inverse relationship between dose of the drug and mean time of response is progressive within a breed although bodyweight clearly exerts an influence. In contrast to the situation in cattle, retention of the placenta is not found, nor is lactation depressed when judged by the rate of lamb growth. Even so, lambs of low birthweight that have probably been dropped several days prematurely need to be carefully watched.

Induced lambing may be seen as a useful follow-up to synchronised breeding procedures but, in practice, there is one significant problem – that of mismothering during the very condensed period of birth. To avoid this pitfall, a large number of lambing pens becomes essential. If a farmer has the necessary labour and facilities, then there is no reason why control of lambing should not become the final step in regulation of the breeding cycle. Control of the date of lambing some 5½ months in advance by means of synchronised oestrus and induced parturition must surely be attractive in large-scale systems of production.

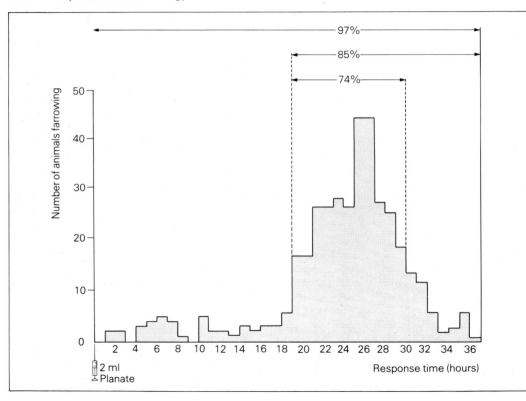

Figure 6.17 The time of parturition induced in pigs by means of a prostaglandin analogue (Planate). When injected shortly before the spontaneous onset of farrowing, quite precise control is obtained over the delivery process

Out-of-season and prepuberal breeding

Out-of-season breeding clearly implies a technology applicable to those breeds of sheep that show a period of anoestrus. Such anoestrus is a reflection of quiescent ovaries, in turn due to lack of stimulation by gonadotrophic hormones. The basic cause of this refractoriness is the changing pattern of daylength, which can be overcome successfully but expensively by housing sheep in lightproof pens and subjecting them to progressively decreasing hours of daylight. This simulates one component of the shift from summer to autumn and, as a consequence, induces oestrous cycles and initiation or resumption of fertility. A simpler, cheaper and more direct approach is to inject the ewes with some form of gonadotrophic hormone, usually in conjunction with a period of ovarian steroid administration via intra-vaginal sponges.

An observation from many trials is that fertility is poorest when ewes are treated in deep anoestrus, whereas it improves progressively as the onset of the breeding season approaches. The timing of a treatment is therefore a matter of judgement, but it may be more realistic to think in terms of advancing the breeding season by 5–6 weeks rather than seeking the objective of year-round breeding. The latter may more sensibly be reached by changing to breeds that have prolonged seasons or do not show a marked anoestrus such as the Dorset Horn, Finnish Landrace or their crosses.

A hormonal approach that has been used successfully in the UK, Ireland and France involves treatment of mature ewes for 14 days with intra-vaginal sponges containing a solution of synthetic progesterone. The role of progesterone is to prime the brain (hypothalamus) for the feedback action of ovarian oestrogens that must precede overt oestrus and ovulation. Ovarian follicular stimulation is then given at the time of sponge withdrawal in the form of a PMSG injection (500–800 i.u.), and oestrus generally follows in two days (Fig. 6.18). At least 60 per cent of ewes mated in June or July after such treatment should produce lambs, but an adequate ratio of rams to ewes (e.g. 1 : 10 rather than 1 : 40) is essential for good fertility at the induced heat. In addition, studies in Ireland have shown a beneficial effect of introduction of experienced rams two days after rather than at the sponge removal. Those ewes failing to conceive to the induced oestrus should return spontaneously approximately one cycle length later.

As suggested above, this form of therapy for extending the breeding season should be much more attractive than erecting costly and inflexible lightproof houses; it should also enable three crops of lambs in two years to be a realistic objective.

Prepuberal breeding

The time from birth until puberty represents a significant proportion of the animal's lifespan. If breeding could be induced in animals of prepuberal age, this might have

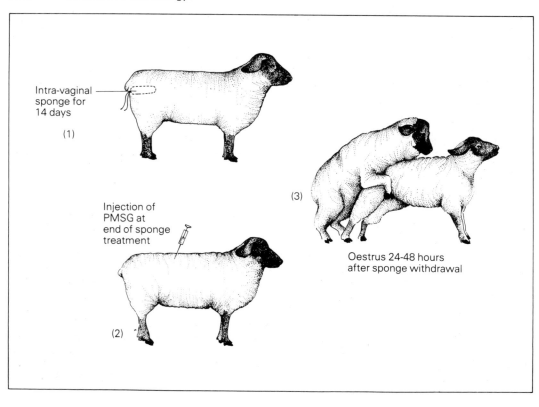

Intra-vaginal sponge for 14 days

(1)

Injection of PMSG at end of sponge treatment

(2)

(3)

Oestrus 24-48 hours after sponge withdrawal

Figure 6.18 Typical scheme for out-of-season breeding of sheep, illustrating the combined treatment with progesterone and PMSG used to achieve this objective. Artificial insemination of such ewes is likely to be the sequel

attractions and it would certainly shorten the *generation interval*. This statement does not necessarily infer that treated females would be expected to carry a pregnancy, for the technique of embryo transplantation could be used after insemination of prepuberal animals that had been induced to ovulate. The relevance of such ideas to commercial farming remains speculative, but experimental work has shown them to be feasible.

As pointed out in Chapter 1, the ovaries are already fully stocked with egg cells by the time of birth, and many of these are within vesicular (Graafian) follicles by two months of age. In this condition, the follicle will respond to a classical treatment with gonadotrophic hormones, i.e. a phase of growth stimulation under the influence of FSH (PMSG) followed 3–4 days later by an ovulating dose of LH (HCG). More complex treatments involve progesterone therapy before gonadotrophic stimulation. Induction of ovulation in this manner should not present problems although, as always, individual responses are unpredictable.

On physical grounds, prepuberal animals are unlikely to support a mature male for mating, so breeding will involve artificial insemination. This may be quite successful in terms of yielding fertilised eggs and embryos, as has been demonstrated in lambs and pigs. Fertility after insemination has been more of a problem in calves, where sperm transport to the oviducts is less efficient and sperm survival is much diminished.

Analogous to the statement already made concerning extra-seasonal breeding, the nearer the animal is in development to puberty, the more successful a prepuberal breeding treatment is likely to be. This is certainly the case where maintenance of the pregnancy is concerned, because, putting considerations of body size on one side, the hormonal mechanisms for supporting the pregnancy (i.e. the luteotrophic mechanisms) only become fully functional shortly before puberty. It is possible to maintain a precocious pregnancy artifically by means of injections of a progesterone solution, but again this is unlikely to appeal in conventional systems of farming.

Selection techniques and sex predetermination

Information from artificial insemination centres based on several thousand or more recordings indicates differences in conception rate attributable to specific males. In fact, even between bulls of supposedly high fertility, one can demonstrate consistent differences in fertility. Farmers therefore need to make a balanced judgement when nominating a sire between his desirable production traits and the expected conception rate. Although results derived from insemination in a range of herds are the most meaningful, individual bulls can be ranked experimentally by the technique of heterospermic insemination. This involves mixing two ejaculates from males whose offspring can subsequently be identified, and noting which male has a consistent advantage in the test. In this way, bulls can be ranked in a hierarchy based on their success in a competitive fertilisation situation. It remains to be convincingly demonstrated that semen from the most 'competitive' bull yields the highest conception rate in a non-competitive situation.

Selection of females for high prolificacy has traditionally been based on performance as judged, for example, by the number of viable lambs or piglets born. However, recent experiments have suggested that it may be possible to predict fertility in prepuberal females – and, to a lesser extent, in prepuberal males – with respect to future daughters by measuring the concentration of gonadotrophic and/or gonadal hormones in their blood. This seems a reasonable premise since the extent of ovarian stimulation and response should be revealed in the secretion of these hormones. A snag in measuring blood concentrations of hormones is that the pulsatile secretion of pituitary hormones in particular requires a very considerable number of samples before meaningful curves can be plotted. While measurement of hormones can therefore be shown to be valuable as a selection tool in research institutes, this approach is again unlikely to feature prominently on commercial farms in the near future.

Sex predetermination

A long-standing objective of reproductive physiologists has been to separate X-bearing and Y-bearing spermatozoa so that insemination of the fractions would lead to offspring of known sex. There have been many claims of success, and certainly the sex ratio has periodically been displaced from approximately 50 : 50 males to females. However, no methodology has yet yielded a predictable, repeatable, and significant shift in the sex ratio, nor at present are there rational grounds for supposing that this can be achieved. Spermatozoa are haploid cells possessing only half the usual number of chromosomes for the species, and there is no evidence that haploid – as distinct from diploid – cells express their genetic constitution in surface characteristics. This being the case, it is difficult to imagine how chemical, electrical or biophysical separation of X and Y spermatozoa might be obtained. One physical difference between the two types is a minute difference in mass attributable to the smaller size of the Y compared with the X chromosome, but this is almost certainly masked by the complement of autosomes plus the mass of the rest of the cell. Even so, centrifugation of semen samples has been one of the most frequent approaches to attempted separation.

In a recent review, the distinguished Edinburgh reproductive biologist, Professor R. V. Short, expressed the opinion that regulation of the primary sex ratio remains a mirage on the horizon. This is not to infer that some quite fortuitous observation will not, in due course, yield a methodology for sperm separation. After all, the use of glycerol as a protective agent for deep-freezing of fowl and bull spermatozoa came as a result of a laboratory accident – in fact a mislabelling of bottles, followed by good detective work by Dr C. Polge and the late Dr Audrey Smith.

The sex of offspring can be predetermined to the extent that placental cells can be stained to reveal the sex of the foetus – having aspirated a small droplet of cells and fluid by the procedures for blastocysts could lead to storage banks of choice of sex is possible if one is prepared to abort foetuses of the unwanted sex, but clearly this is wasteful, has emotional overtones, is not applicable to litter-bearing species, and in no sense influences the genetic sex of a subsequent conception. Amniocentesis is likely, therefore, to have a negligible application in farm animals. On the other hand, current experimental studies on determining the sex of very young embryos – blastocysts of approximately one week of age – prior to transplantation into a recipient might have a role in certain commercial operations. The procedure here is to snip a few cells from the wall of the blastocyst (a manoeuvre requiring a micromanipulator), stain them for the presence or absence of sex chromatin to determine whether they are from an XX or XY embryo, culture the blastocyst for a few hours until the puncture wound is healed, and then carry out transfers of embryos of known sex. This has been shown to be perfectly feasible in rabbits and mice, and with deep-freezing procedures for blastocysts could lead to storage banks of embryos of known sex.

In the near future, therefore, not only will transplantation of embryos of known genetic background be widespread, but at the same time it may be possible to specify the sex of the offspring.

Further reading

Hunter, R. H. F. (1980) *Physiology and Technology of Reproduction in Female Domestic Animals*, Academic Press, London and New York.

Hafez, E. S. E. (ed.) (1980) *Reproduction in Farm Animals* (4th edn), Lea & Febiger, Philadelphia.

Cole, H. H. and Cupps, P. T. (eds) (1977) *Reproduction in Domestic Animals* (3rd edn), Academic Press, New York and London.

Nalbandov, A. V. (1976) *Reproductive Physiology of Mammals and Birds* (3rd edn), Freeman, San Francisco and London.

Austin, C. R. and Short, R. V. (eds) (1972–1976) *Reproduction in Mammals* (espec. Vols 1–6), Cambridge University Press, Cambridge.

Glossary of abbreviations

ACTH	Adrenocorticotrophic hormone
BSA	Bovine serum albumin
CL	Corpus luteum (singular); corpora lutea (plural)
DMSO	Dimethyl sulphoxide
FGA	Fluorogestone acetate
FSH	Follicle stimulating hormone
FSH-RH	Follicle stimulating hormone releasing hormone
Gn-RH	Gonadotrophin releasing hormone
HCG	Human chorionic gonadotrophin
i.d.	Inside (internal) diameter
i.u.	International units
LH	Luteinising hormone
LH-RH	Luteinising hormone releasing hormone
o.d.	Outside diameter
$PGF_{2\alpha}$	Prostaglandin $F_{2\alpha}$
PMSG	Pregnant mare serum gonadotrophin
RF	Releasing factor
RH	Releasing hormone
TSH	Thyroid stimulating hormone (thyrotrophin)
w/v	Ratio of weight/volume

Units

mg	milligram (10^{-3} g)
μg	microgram (10^{-6} g)
ng	nanogram (10^{-9} g)
pg	picogram (10^{-12} g)

Glossary of scientific terms

Most technical terms are defined in the text as they arise, but there follows a list of selected words that may prove useful when perusing the book at random.

Acrosome — Sac-like organelle on anterior portion of sperm head, containing enzymes used during penetration of egg membranes.

Ambient — Surrounding (i.e. environmental) with respect to temperature.

Amniocentesis — Technique of aspirating placental fluid in order to check chromosome constitution of foetus.

Analogue — Synthetic chemical derivative, usually having biological properties of a hormone or hormonal-like substance.

Anoestrus — Phase in non-pregnant animals in which oestrous cycles are suspended, due to influence of lactation, season and/or poor nutrition.

Atresia — Degeneration and wastage of ovarian oocytes and follicles.

Blastocyst — Embryo of a few days old in which a fluid-filled cavity (the blastocoele) has been formed by rearrangement of the cells.

Capacitation — Final maturation of spermatozoa in the female reproductive tract enabling fertilisation.

Caruncle | Maternal component of placenta in ruminants, specialised for attachment to the foetal cotyledon.

Chemotaxis | Phenomenon of attraction between cells, especially attraction of motile spermatozoa by eggs or their investments.

Colostrum | First milk of lactation containing a high concentration of antibodies (immunoglobulins).

Conception rate | Number of animals pregnant expressed as a percentage of the total number mated or inseminated.

Conceptus | An expanded and elongated blastocyst – the condition of the embryo in farm animals shortly before attachment.

Cotyledon | Foetal component of placenta in ruminants, specialised for attachment to the maternal caruncle.

Cryptorchid | Condition in which testes remain in abdominal cavity instead of descending into scrotal sac.

Dystocia | Difficulty at the time of birth, especially of delivering the foetus through the pelvic canal.

Endogenous | Originating within the animal, particularly hormones (cf. Exogenous).

Endometrium | The glandular epithelium (internal lining) of the uterus.

Exogenous | Originating outside the animal and introduced, for example, by injection.

Freemartin | Female (especially calf) that has been co-twinned to a male, and is infertile due to abnormalities of the reproductive tract and/or gonad.

Fremitus | Change in pulsation character ('vibration') of an artery upon alteration of digital pressure.

Gonadotrophic | Refers to hormone(s) that stimulates the gonads, especially maturation of ovarian follicles.

Gravid | Condition of the uterus (i.e. containing an embryo or foetus) when an animal is pregnant.

Haemotrophe | Blood borne sustenance to the embryo, obtained via the placenta.

Histotrophe | The fluid ('uterine milk') composed of uterine secretions and cellular debris that sustains the embryo before attachment.

Hyperplasia | An increase in cell number.

Hypertrophy | An increase in cell size.

Hypophysectomy | Surgical removal of the pituitary gland (i.e. the hypophysis).

Hypoplasia | Retarded or reduced cell division, giving deficient growth.

Hypothermia | Condition of excessive loss of heat giving reduced body temperature.

Hysterectomy	Surgical removal of the uterus.
Involution	Restoration of uterus to its cyclic condition after pregnancy.
Karyotype	Technique of preparing chromosome spreads to check the genetic constitution – usually of diploid cells.
Laparoscope	Form of telescope (endoscope) introduced into the body cavity for visualising the ovaries and/or uterus.
Libido	Sexual drive, especially of males.
Lordosis	Behavioural stance in oestrous females, usually involving arching of the back and 'pricking' of the ears.
Luteolysin	(Adjective: luteolytic) Substance acting to reduce or terminate progesterone secretion by the corpus luteum.
Luteotrophin	(Adjective: luteotrophic) Substance that stimulates progesterone secretion by the corpus luteum.
Monotocous	Production of single offspring per pregnancy (cf. Polytocous).
Motility	Phenomenon of spontaneous movement in cells, i.e. swimming activity of spermatozoa.
Necrosis	Death of cells or tissues.
Neonatal	Shortly after the time of birth; pertaining to the newborn.
Nymphomania	Persistent and intense oestrus (heat), particularly in cows, usually with associated bellowing noise.
Oedema	Tissue swelling due to infiltration of lymph into the mucosal layers.
Oestrus	Behavioural condition of female receptive to mating. Synonymous with 'heat' or 'season'.
Ovariectomy	Surgical removal of ovaries.
Ovulation rate	Number of eggs ovulated per oestrous cycle.
Parity	Number of previous pregnancies or litters.
Parous	Condition of already having had a successful pregnancy.
Parturition	The act or process of birth.
Perinatal	Just before, during, or just after the time of birth.
Peritonitis	Inflammation and/or infection of the peritoneal (body) cavity.
Pheromone	Odoriferous substance given out by one animal that acts as a signal to another of the same species.
Placentome	Attachment complex in the placenta of ruminants formed by interdigitation of foetal cotyledon and maternal caruncle.
Polyspermy	Penetration of egg cytoplasm by two or more spermatozoa, invariably a lethal condition in mammals.

Polytocous	Producing several (or more) young at birth.
Prenatal	Before the time of birth.
Progestagen	Synthetic form of progesterone.
Pyometritis	Infectious pus in lumen of the uterus giving temporary or prolonged infertility.
Silastic	Form of silicone rubber.
Superovulation	Increase in ovulation rate over that characteristic for the species, usually by hormonal treatment.
Surfactant	Protein secreted in lungs of foetus shortly before birth to act against surface tension and prevent collapse of lung.
Syndrome	Condition of two or more concomitant symptoms.
Synergise	Acting together to promote an effect.
Trophoblast	Cells of the blastocyst that proliferate to form the embryonic membranes, notably the chorion.
Vasectomy	Rendering male sterile by cutting each vas deferens (spermatic cord).

Index